Nelson Guide to

Report Writing

Nelson Guide to

Report Writing

Lawrence Gulston

Sir Sandford Fleming College

tralia Canada Mexico Singapore Spain United Kingdom United States

Nelson Guide to Report Writing

by Lawrence Gulston

Editorial Director and Publisher:
Evelyn Veitch

Executive Editor:
Anne Williams

Marketing Manager:
Lisa Rahn

Publisher's Representative:
Bill More

Senior Developmental Editor:
Mike Thompson

Executive Production Editor:
Susan Calvert

Copy Editor/Proofreader:
Kelli Howey

Indexer:
Edwin Durbin

Production Coordinator:
Ferial Suleman

Creative Director:
Angela Cluer

Interior Design:
Gabriel Sierra

Cover Design:
Johanna Liburd

Cover Image:
© Corel

Compositor:
Kathy Karakasidis

Printer:
Transcontinental

Printed and bound in Canada
1 2 3 4 07 06 05 04

For more information contact Nelson, 1120 Birchmount Road, Toronto, Ontario, M1K 5G4. Or you can visit our Internet site at http://www.nelson.com

National Library of Canada Cataloguing in Publication Data

Gulston, Lawrence
Nelson guide to report writing / Lawrence Gulston.

Includes index.
ISBN 0-17-641416-9

1. Technical writing. I. Title.

T11.G86 2004 808'.0666
C2004-900071-3

CONTENTS

PREFACE

We have designed the *Nelson Guide to Report Writing* to be a brief, useful guide to current report format and writing practice for scientific and technical writers at college and in the workplace. A report writer can use it as a course text, lab manual, or writer's guide and resource as he or she develops a piece of scientific or technical writing of any length. This guide directs students to deal with the central issues of writing a report:

- researching and organizing information
- writing in a clear, readable style
- formatting report pages
- dividing the text into sections
- writing appropriate report elements
- integrating graphic presentations to support text descriptions
- documenting secondary sources of information

The student writer typically consults a book like this after attempting a report assignment and finding trouble. But reading this book *before* writing that first draft of your report will speed up the writing process by reducing the amount of revision necessary. You will be able to start the first draft with the right elements in the right place, the right graphics in the right place, and the sources all cited and listed in the correct form. Starting off right means less work and better results—and, in addition to all this, you will be more confident as you begin your report.

This book is a guide to the fundamentals of report writing. Each topic discussed is an overview of a complex subject, from documentation systems to styles of rhetoric to grammar and sentence syntax. We hope that the inclusion of frequent specific examples of scientific and technical reporting will help students and professional writers produce documents that communicate effectively. For those requiring greater detail or a more theoretical approach, textbooks are available that cover composition and technical writing specifically.

Professors who assign reports will have specific requirements for their students' writing projects. It is important for faculty to set out format expectations for written work in clearly stated assignments.

Writing about scientific and technical work is the best way for the student or professional scientist or technician to review, evaluate, refine, and ultimately share his or her work. Clear, concise writing requires the writer to think clearly and concisely about the issues of evaluating scientific and technical literature, keeping laboratory and field records, describing and interpreting scientific and technical data, and communicating data and analysis to various audiences in a report.

The report format recommended in this guide is primarily for scientific and technical reporting. It includes an abstract, used in place of the executive summary in a scientific paper. General business reports differ in significant ways from scientific and technical reports, and they vary considerably in design and in content.

At the suggestion of reviewers, the organization of this book reflects the writing process. Students must consciously follow a sound methodology in constructing a technical or scientific document in order to achieve efficient, effective communication. Their tendency to focus on details of format inhibits their ability to focus on the overall goals of technical communication and appropriate methods of achieving those ends. We do present information about report format as a reference for students with specific questions or issues to resolve. However, they must always see this problem-solving activity in the larger context of the writing process.

The section on documentation describes the three major formats currently in use in scientific and technical reporting. Writers and professors will be able to choose the format that best suits their needs. Examples of each format cover electronic sources currently in widespread use, such as Web pages, CD-ROMs, and e-mail messages.

The Appendix elements feature a sample student report and tutorials for creating specific report formatting using Microsoft Word XP™ word-processing software. Methods are provided for creating reference lists, inserting mathematical equations, and wrapping text around graphic figures.

Acknowledgments

I wish to acknowledge the support, encouragement, and patience of the Nelson Thomson publishing team: Bill More, Anne Williams, Mike Thompson, and Susan Calvert. I would also be remiss if I did not acknowledge the support of the faculty, staff, and students at Sir Sandford Fleming College, without whom this first edition would not have been written. Finally, I wish to dedicate this book to my children: Jennifer, Sandy, Brian, and Anthony.

The author and the editors would like to thank the following people who reviewed the manuscript and provided their valuable insights and suggestions: Stephen Boaro, Northern College; Heidi Janes, Marine Institute of Memorial University; and Fleurette Simmonds, Durham College.

Chapter 1

Report Research and Organization

Overview: This chapter addresses the most important part of report writing: identifying your audience and purpose. It also discusses team-based report writing, effective research methods, organizational processes, and planning.

AUDIENCE AND PURPOSE

Audience

Have you ever found yourself sitting in front of a blank computer screen when you have to write a report, wondering what to write about, or even how the document is supposed to look? Did you know that there is a simple answer to those questions?

Everything you need to know about content and format in your report can be found by looking at the needs of your intended audience and your purpose in writing. Thinking about these things before you write will save you time and help you produce a better report. You may ask yourself, "Should I tell the boss in my report about the late delivery on Thursday afternoon?" Your answer could conceivably be, "No—it didn't put the project behind schedule and that's what matters to my boss, who has to read my report."

We often write things down to satisfy personal needs and goals, like a diary or an e-mail to a friend. However, the words we use sometimes do not signal all of our intended meanings. Words can be incomplete expressions that our readers need to add to with their own knowledge base, or from their knowledge of our personality. When we reshape our words into complete meanings in order to satisfy the needs and goals of others, we begin to communicate efficiently and effectively. Good communication, then, gives both personal satisfaction and professional success.

The reshaping process takes work. A scientist or technician can expect to spend 20 percent of his or her working week creating written documents. That's one working day out of five. The goal of learning good technical communication is to make that writing time efficient and effective.

The Goals of Technical Writing

The purpose of scientific and technical writing is to present your readers with accurate, complete, and current information about your work that is appropriate to their needs. Conventional types of scientific and technical reports have structures that meet the needs of readers in specific workplace situations. Workplace reports include the progress report that is sent to update managers on the status of company projects, and the proposal that describes potential company projects in detail and persuades the reader to approve the work.

Scientific and technical data are often complex. Writing about them requires that you organize your workplace data and concepts into a coherent narrative that provides understanding and accessibility for your readers.

The Readers of Scientific and Technical Writing

The primary audience for your report will be the person or group of people who will read your document and take responsibility for its contents, such as following up on recommended actions. You may have a secondary audience: persons or groups of people who are interested in its contents, such as technicians doing similar work, company executives monitoring progress, and so on.

The better you know your audience, the better you will be able to know what they want to read about and in what form. Personal communication with your audience is the best way of finding complete information about an audience's needs, from asking your boss which field results to include in the monthly report, to distributing a survey among park visitors to determine the topics most interesting to them for a park newspaper or outdoor presentation, to confirming the level of technical knowledge company executives possess before presenting them with a report filled with terms like megahertz and gigabytes. However, experience and some general guidelines will allow you to anticipate your audience and their interests.

Types of Audiences

Readers of scientific and technical literature can be grouped according to similar interests as follows:

- *Experts* are well educated and look for scientific and technical data, usually presented in graphic form, and for interpretation of experimental results.
- *Technicians* look for solutions to practical problems.
- *Managers* look for information about how time and money were spent and for cost-effective solutions to technical problems.
- *Operators of equipment* look for clear, well-illustrated instructions.
- *The general public* looks for simplified explanations of processes and products.

The Purpose of Technical Writing

Most scientific and technical reports are written to convey specific information to specific readers. The word "report" itself comes from two Latin words meaning "to carry back." The most important part of your job, once the lab or fieldwork is done, will be to "carry back" the results to people in your organization who need the information.

As you become proficient in writing reports, you will find yourself better able to influence workplace decisions. Supervisors will ask you to follow up on well-researched and organized reports with recommendations for action in specific situations. These reports have a persuasive purpose, in addition to their basic function of informing the reader.

Other goals of scientific and technical writing include the following:

- to uncover facts
- to separate fact from opinion
- to make valid and convincing conclusions
- to record and discuss your findings from field and lab work
- to provide facts that support an informed opinion
- to move from general data, such as is found in encyclopaedias, to specific information, such as is found in reports, field studies, textbooks, and monographs

State your purposes clearly in the introduction to your report. Use action verbs: "to investigate," "to develop," "to compare and recommend," and so on. Your reader wants to know why your report was written. Writing down your purpose helps you clarify your work goals and what to include in the report. The report at the end of your lab or field investigations brings back to your workplace—and your supervisor—the information that is useful, often vital, to the company's operation. Clearly written and logically presented, the data in your report are the key to your career in science and technology.

WRITING IN TEAMS

Purpose

Writing in teams, or *collaborative writing*, is common in the workplace because it has the following advantages:

- *Expertise*—To reach a common goal successfully, scientific and technical projects require a wider range of skills and expertise than a single writer can provide.
- *Feedback*—A writing team can provide a variety of viewpoints on the developing text in terms of its style and its content. This improved feedback in most cases gives a report that better communicates with its intended audience.
- *Networking*—Scientists and technicians find that working on writing projects in teams is less lonely and provides important opportunities to network and share information. It improves communication among employees and provides an introduction to corporate culture for new employees.

However, writing in teams has the following disadvantages:

- *Time*—Collaboration means communication, which takes time. Writing teams require more time to produce documents than individual writers.
- *Conformity*—Team members often work harder at co-operating with each other than at asking difficult questions of each other in order to produce a better document.

- *Workloads*—In most teams, some members do more work than others. Some team members remain unmotivated to contribute work of good quality because they have been given less to do, or because they perceive they have too much to do, or because they have experienced conflicts with other team members.

Team writing will not go away because of these potential problems. In the modern workplace, the ability to work with others has become as important as technical competence. A wise strategy for a college student is to learn better interpersonal and communication skills to complement their scientific and technical skills.

An Approach to Team Writing

Two basic approaches, co-authoring and consulting, are commonly used by workplace writing teams. *Co-authoring* in teams of two or three people means that each team member takes responsibility for organizing and writing a section of the text, with regular feedback from the other team members. *Consulting* in larger teams means that a writer or writing team takes responsibility for developing the plan of the report, co-ordinating the expert input of the other team members, and editing the text.

Members of successful writing teams are aware of the interpersonal dimensions of the team and are willing to take personal responsibility for their actions. They treat the other team members as they themselves wish to be treated. They respect the other team members' strengths, seeking out their ideas and special skills and finding ways to use them in the writing project. They insist on excellence in their own work and in the project's final documents, but without becoming perfectionists regarding unimportant details.

The interpersonal skills required of a good team writer are the following:

- conducts efficient, diplomatic team meetings
- listens to others and contributes ideas
- asks good questions
- uses technology effectively to communicate

RESEARCH

Gathering Information

Before you write, you must have something to write about. Finding a topic is usually not a problem in the workplace, as circumstances will make your subject matter clear. Your focus, then, must be on getting all the information about the situation and selecting details carefully for your chosen reader or readers. These details will come from either primary or secondary sources of information; the following section distinguishes these two sources and shows you how to generate information from each.

Primary Sources

Your own training and ability in lab and field procedures will be reflected in the notes you take. Recording accurate data on a clipboard or personal digital assistant (PDA) means careful attention to scientific detail and technical processes. A good technician double-checks details and figures for accuracy. Accurate detail is the lifeblood of technical reporting. A scientist must record every detail of a procedure, even the mistakes. Here are the main points to keep in mind:

- Record experimental information in a notebook with bound pages.
- Open each entry with the date and time.
- Use the same format for similar kinds of experimental procedures.
- Record everything that happens, but avoid explaining anything.
- Avoid jumping to conclusions: the proper place for interpreting your data is in the final report. Do not confuse inferences and judgments with facts.
- Use standard format on data sheets or readouts from instrument faces for most field and lab recording. Use your notebook to record extra information not included on standard forms and instrumentation, such as weather conditions, unusual variations in meter readings, and so on.
- Write down observations as soon as possible after they happen. Such timely journal entries will be essential to accurate interpretation of the data later on.

These lab and field data, as well as the overview of the project from your personal notes, are your primary sources of information and form the basis of the text of your report, as well as the source materials for its tables and figures. Good notes will make a good report.

Secondary Sources

In most reporting situations, you will be asked to interpret the data you have collected. In order to draw valid inferences and support your conclusions with authority, it is necessary to read scientific and technical material written by others in professional journals, technical bulletins, conference papers, and the like, and then to compare these published findings with your own data.

Secondary sources are also widely scanned by technical staff for solutions to technical problems. If, for example, an engineering technologist is faced with unacceptably high current drains in a new electro-mechanical device being designed, he or she will go looking for ways to reduce the current drain while maintaining optimum output performance by researching similar designs in recent technical literature.

Finding Secondary Sources

Library and Web research are the most common sources of scientific and technical information for college and workplace reports. Company and department files also supply timely, specific information for reports. Prioritize your research by looking in the most likely places first for the information you and your reader need.

Libraries

A library is organized for easy access of materials and contains information in print. Most North American libraries use the Library of Congress classifications, although some local libraries still use the Dewey decimal system. Know the system in your library so that you can locate useful materials quickly, and learn to use reference tools, such as subject indexes and specialized dictionaries and directories. Learn about library resources during each trip you make there. Information in print has been reviewed and edited during the publishing process and selected by library staff to meet readers' needs, making library resources generally more reliable, accurate, and thorough than any other.

The Internet

The Internet is not a library. It is a dynamic system of billions of electronic documents and digital files, each with a very short average lifespan on the Web. The public Internet is accessible by using browser software and search engines like Google, directories like Yahoo, and metasearch engines like Metacrawler. Learn how to select key words carefully, observing spellings and shades of meaning. A small change in spelling can give you a big change in search results. Conversely, the private Internet (or *intranet*) consists of files stored on local servers that cannot be found by search-engine software following hyperlinks. Intranets are usually reserved for corporate use only. All files can be located by the Uniform Resource Locator (URL) system, often called the "Web address." Most Web-wise researchers look for moderated lists of links in their subject areas. Experts in various scientific and technical fields will search the Internet for Web sites with current, authoritative, complete information and put up pages of hyperlinks to these resources. Each link will have a summary and assessment of the information on the linked Web site. Since so much information on the Web is incomplete or inaccurate, such lists of good resources by knowledgeable people can be quite useful, but they must be current. Always check pages for the date they were last updated.

The Internet can also help during the drafting phase of writing, when you can access online dictionaries, thesauruses, encyclopaedias, and writing labs that deal with matters of report style and format. It is also a source of digital image files to support scientific and technical descriptions in your text.

Using Secondary Sources

As you read your source information, write paraphrases of the key ideas and information. Paraphrases are your own personal combinations of the key words and phrases of the original. Paraphrasing involves distilling what you select in your reading as important information, and then writing it down as a collection of points.

Beware the photocopier and the Web page. They make it easy to ignore the essential process of reading text and understanding ideas. Is that hard work? Yes! Admittedly, it is easier to highlight some phrases in a photocopy or to copy and paste a section of text from a Web page to a draft in your word processor. But you still won't understand the original completely, and your final report will still sound like somebody else wrote it. If you did not cite your source of information, the disaster is complete; you've committed plagiarism.

Before taking notes from secondary sources, write down in your notes all the details about your source of information that are required by your documentation system. You will need these later for citations and reference-list entries in your report.

When opting to use quotations, keep the following points in mind:

- Select brief quotations carefully;
- Use quotations for specific examples of general ideas and for exact statements of general principles;
- Limit direct quotations to one or two sentences wherever possible;
- Limit direct quotations to two or three examples per report.

ORGANIZATION

The Importance of Organizing

Finish your research when your project has ended and no data are left to report, or when time runs out before a reporting deadline. Leave yourself enough time before the report deadline to organize the information, write and revise the drafts, and set up the document in its final format.

Organizing your research is the key to readability in your report. Use ideas to organize facts, and learn common patterns of organization and let them guide your report plan: comparison, more- to less-important, classification and division, cause and effect, problems/solutions, and so on.

Look for ideas in your notes or in discussions with other professionals. If you have a visual imagination, try sketching your ideas. Try to answer the time-honoured journalistic questions: who, what, when, where, why, and how. Your own knowledge and experience will suggest relationships within your research information. To make sense to your reader, your facts must be connected to each other and form a pattern.

Outlining

Outlining is the process of grouping your information together into logical units and making descriptive headings for those units. Planning before you write saves you time and gives your report the quality of structure, a logical progression of facts that helps your reader understand the material faster and better.

A formal outline is an information hierarchy. It consists of descriptive headings that you will place in the report to inform your reader of the location and the type of information presented throughout the report. When you write the first draft of your report, the headings will guide you as you develop the text in each topic area that you have to cover.

Outlining takes practice. Fortunately, you will have many opportunities in college and in the workplace to practise the art, starting with routine e-mails and moving on to longer forms of writing. Do not miss an opportunity to organize a piece of writing, no matter how small. The following is a simple, five-step method for working with your research notes to produce a formal outline of headings.

1. Collect your research notes and read them through until you know them completely and the big picture begins to emerge.

2. Divide your notes into two or three or four major areas of concern. Give each of those areas a name. Use the key words from your research notes to form the descriptive title. Do not worry about details here. If something does not fit into your main categories, set it aside. Consider it later; it will usually fit into a more completely developed outline, or it will appear irrelevant to your report and can be eliminated.

3. Look at each major report topic in turn. Divide each into two or three or four subtopic areas, and give each a more specific topic heading. A topic heading consists of a noun and its modifiers. Continue this process, dividing this second level of organization into a third level with headings, and so on until you find you can write several paragraphs under each heading.

4. Give the outline a working title using the key words in your main headings, and check your outline for a logical sequence of headings and for errors such as faulty co-ordination, faulty subordination,

or faulty parallelism. *Co-ordination* means placing equally important headings at the same level of organization. *Subordination* means placing information in the category where it logically belongs. *Parallelism* means using the same key words in descriptive headings for similar kinds of information to signal the similarity clearly to your reader and make the comparison of your data easier.

5. Format your outline in a standard form, using either the traditional numbering system or the decimal numbering system for the headings. Indent each level of organization about five spaces. The format for the formal outline allows you to see more clearly, through numbering and indentation, the relationship of the ideas expressed by the headings. This makes a final check for errors easier.

Outlining requires taking the time to become familiar with the research information for your report and to work through the development of a hierarchy of descriptive headings. However, it saves time in the long run by speeding up the process of drafting the report. It also produces a better-organized product. Documents describing scientific and technical topics need to have a simple, clear plan of organization. That plan must also be available to the reader so that he or she can find specific topics to meet specific needs in a timely manner. For this reason, your outline in your report will form the basis of your table of contents (see Box 1.1).

When you have finished your outline, check it for errors, a formal numbering system, and patterns of organization. Good thinking skills are important in constructing a formal outline because the hierarchy of headings depends on logical connections. There are many practical ways to improve your thinking skills. Applying the CoRT Thinking Skills© developed by Edward de Bono is a good example. Taking a college-level course in thinking skills is another. A good technical writer is also a good thinker, working through technical data and ideas methodically and logically according to sound thinking principles. This process will be reflected in the structure of headings in a report outline.

Common Patterns of Organization

Organizing your information is your first and most important step in connecting your report with your audience. Keeping the following points in mind will help you to organize effectively:

- I. Introduction
  - A. Purpose and Scope
  - B. Review of Literature
  - C. Study Area
  - D. Background

- II. Maya Writing
  - A. Sources of Texts
    - 1) The Codices
    - 2) The Stelae
    - 3) Monuments
    - 4) Ceramics

  - B. Forms of Maya Writing
    - 1) Letters
    - 2) Numbers
    - 3) Phonemes

- III. The History of Maya Epigraphy
  - A. Tatiana Proskouriakoff
  - B. Linda Schele
  - C. David Stuart and the New Epigraphy

- IV. Glyphs of Three Major Cities
  - A. Tikal
  - B. Copan
  - C. Palenque

- V. Conclusions
  - A. Altering Maya History
  - B. Importance for Modern Maya
  - C. Importance for Archaeology

Box 1.1: Outline for Report on Maya Epigraphy

- Learn to recognize common patterns of organization in reports, each one a flexible sequence of topics in a logical order, and use them. Your reader will also recognize them and thereby understand the content better and faster.

- Use patterns of organization flexibly. They will guide your thinking as you look for relationships in the data contained in your research notes. These patterns are not rigid rules or formulated categories, like a questionnaire or lab sheet; they are patterns of thinking that form the basis of much scientific thought. As such, they give a recognizable structure to your writing and provide the basis for drawing logical conclusions from your data.

- Modify and combine these patterns to suit your report. If you need to explain one aspect in greater detail than the others, do so. Make sure your descriptive headings reflect this amount of detail by showing more subheadings in this area.

- Combine patterns in longer reports by using different patterns to organize different material within major and minor sections of the report. For example, in a report comparing three or four computer systems, set up criteria for comparing them first, listing and describing the criteria in the more- to less-important pattern.

Some patterns of organization are so simple that they need no explanation. Time and space are two of these. Describe, for example, technical procedures step-by-step in the sequence in which they happen in real time. The same is true of technical processes, such as signal processing in a radio receiver, beginning with the radio frequency input. When organizing spatial information derived from fieldwork, move consistently north to south or east to west, describing the features of each area in turn.

Comparison

Using the Pattern

Comparison is common in reports that develop scientific and technical data for two or more possible choices and end by recommending one of them. For example, feasibility studies for new landfill sites require environmental assessment of several possible locations and make recommendations regarding the best choice. The basic pattern of the report is comparing Site A with Site B, and so on.

In individual paragraphs and sentences explaining scientific and technical processes and principles, this pattern allows the writer to expand the reader's understanding from something familiar to something unfamiliar. For example, to explain a zebra to a child who has never seen one, the writer can begin by comparing the zebra to something familiar, such as a horse, to give an idea of size and conformation. Then the writer can add what is different about the zebra, most notably its black-and-white stripes. This pattern of development is sometimes called the "given-new" strategy. Show your reader the similarities between something familiar and the unfamiliar process or principle you are explaining, and then show the differences.

Establishing Criteria

Establish criteria for your comparison when using this pattern for the major sections of the report. Criteria are the needs and standards used to compare data. If criteria are not applied consistently to the data, the comparison and the conclusions drawn from it will be invalid. For example, describing one computer's motherboard with a comparison to fast bus speeds, and then describing another computer's hard drive with a comparison to an Ethernet card does not provide the basis for a valid comparison.

Choosing between Alternative Comparisons

Most writers faced with comparing complex scientific and technical data will have to choose between two possible methods of doing so. It is better to understand the choice and make it knowledgeably. Many writers struggle to express complex detail in writing because they have not grasped the underlying pattern of development.

The two choices are whole-by-whole and part-by-part. In whole-by-whole comparison, the writer describes each alternative in the comparison completely and thoroughly (see Box 1.2; note that only section II of the example outlines appears in the following boxes). The description of each alternative is organized into categories that correspond to the criteria established at the beginning of the report.

Whole-by-whole comparisons work best when the information is relatively simple and straightforward, without complex detail.

In part-by-part comparison, the writer applies each criterion for comparison to each alternative in turn (see Box 1.3). The description of each criterion is organized into categories that correspond to the alternatives being compared.

Part-by-part comparisons work best when information is complex; use this pattern when both the criteria and the alternatives that you are exploring require more detailed description and explanation of scientific and technical principles and processes.

Either of these patterns of comparison may be combined in a report with the more- to less-important pattern. If it applies to your situation, prioritize the criteria for comparison and present them to the reader in more- to less-important order.

More- to Less-Important

Most readers of scientific and technical reports want to know what is important. They rely on your judgment about the factual information and the concepts presented in your report. The *more- to less-important* pattern presents the main ideas first (see Box 1.4).

Look at the points you want to make in a section of your report. Are some more important than others? Can you say why? If the answer to these questions is yes, begin with an explanation of the reasons for your set of priorities. Then organize your points in a list from the most important to the least important.

The more- to less-important pattern requires that you evaluate your data objectively and describe your priorities or your company's priorities fairly. The scientific reader, for example, expects to read the most important research findings first. The executive reader expects to read answers to the most urgent questions first. The more- to less-important pattern is particularly useful in persuasive writing, such as one finds in a business proposal.

II. Comparison of Systems
   A. Dell
      1. Software
      2. Memory
      3. Disk Storage
      4. Cost
   B. IBM
      1. Software
      2. Memory
      3. Disk Storage
      4. Cost

Box 1.2: Whole-by-Whole Comparison of Computer Systems

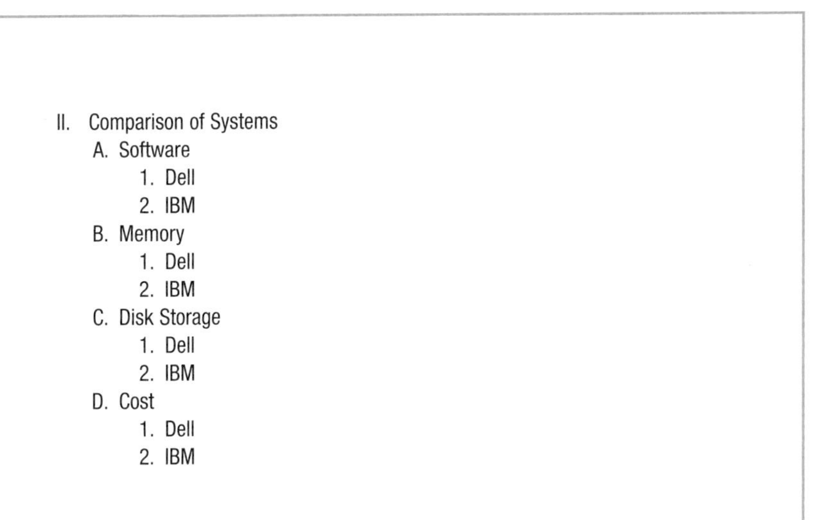

II. Comparison of Systems
    A. Software
        1. Dell
        2. IBM
    B. Memory
        1. Dell
        2. IBM
    C. Disk Storage
        1. Dell
        2. IBM
    D. Cost
        1. Dell
        2. IBM

Box 1.3: Part-by-Part Comparison of Computer Systems

II. Reasons for Choosing My College
    A. Excellent Reputation
    B. Lower Fees and Living Costs
    C. Semi-rural Location

Box 1.4: The More- to Less-Important Pattern

When summarizing the reasons for a particular action, or the benefits of choosing your company to do the work, place your descriptions of the most important reasons or benefits first.

Use vertical lists with numbers or bullets to summarize the main points in a more- to less-important list. Make descriptive headings for your points, and use the headings to show the pattern of organization.

Classification and Division

Classification is the process of putting items with similar characteristics into the same category. In a broad sense, classification is what scientists and engineers do when they gather information about how the world works; they group phenomena, everything from the natural world to building materials, into useful categories.

Induction and Deduction

The thinking process that looks first at the details and then classifies them into larger, descriptive categories is called *induction*, or *inductive reasoning*. Categories help us to organize and understand the world. In a small sense, you can do the same for your reader with the information that you have researched for your report. Categories can also, in turn, contain categories. This is because some ideas are large and inclusive of many observable details, while other ideas are smaller and include less detail, depending on finer and finer distinctions within a basic category.

Division is the process of dividing a general subject into specific component parts. The thinking process that looks first at the main idea or theme and then divides it into smaller, descriptive categories is called *deduction*, or *deductive reasoning*. Use division for descriptions of mechanisms, instruments, and processes. For example, in the description of an instrument, divide the instrument into a number of discrete component parts. Describe each part by dividing it into its component parts and explaining the form and function of each.

A good example of this concept is the trees and the forest. If you start with the trees, instead of the forest, you are classifying, or using induction. You will see that the white pine and the red pine are both pines; that is, they share basic characteristics, along with the jack pine and the Scots pine. The same process of looking at the data will add categories such as firs and spruces. Eventually you may see the white oak and the red oak, giving you oaks to go with maples, birches, and poplars. At the end of the process, you will see that the pines, firs, and spruces share characteristics different from the oaks, maples, birches, and poplars. Thus you now have two groups, softwoods and hardwoods, which are the two largest categories of trees.

If you start with the forest, you are dividing, or using deduction. You will divide the forest into the softwoods and the hardwoods, and then divide those in turn into the various categories of both types. The process ends when you are describing individual tree species, the smallest data set.

The Abstraction Ladder

A way of visualizing classification and division is the *abstraction ladder* (see Figure 1.1). Imagine a household stepladder. At the narrow top of the ladder is the main idea, or the *theme* of your report. At the broad base of the ladder are your data in all their diversity and individuality. When you move up the ladder from the particulars to the general ideas, you are classifying; when you move down the ladder from the general to the particular, you are dividing.

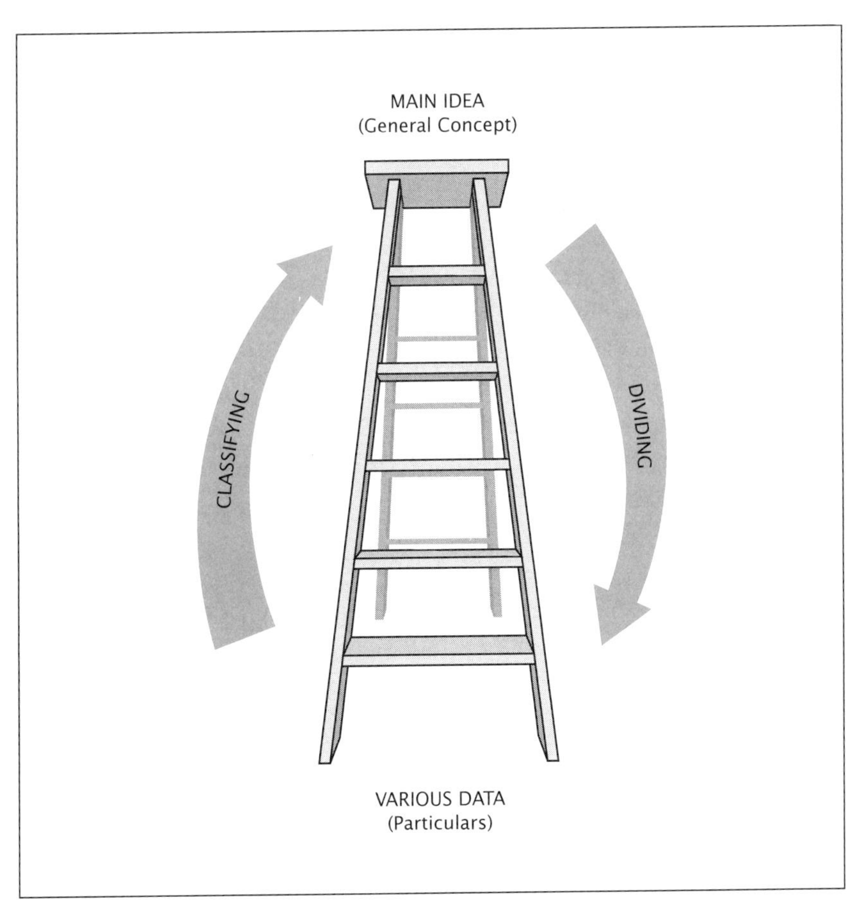

Figure 1.1: Abstraction Ladder

Using Classification and Division

Keep the following points in mind as you develop a classification system for your report.

- *Avoid overlapping categories in your classification system wherever possible.* For example, if you used "water snakes" and "poisonous snakes" as categories to describe reptiles, snakes like the deadly water moccasin would fit into both categories. Work through the logic of categories in your outline to ensure that characteristics forming the basis of classification remain consistent throughout the outline and at each level of the outline. In some areas of science and technology overlap cannot be avoided, because definitions of characteristics are not precise. In these cases, keep overlap to a minimum to provide your reader with a clear, logical classification system.

- *Remember to write for your audience and purpose.* A scientific audience will prefer induction, and expect you to present the data as you found them before reading your analysis and any general conclusions you may wish to draw. A business audience will prefer deduction, and expect you to state the main points of your report first before providing the technical details in logical categories. Check the characteristics on which your categories are based. Will your audience find this distinction useful? For example, categories of canoes based on fine distinctions between regular and ultralight Kevlar construction will suit knowledgeable tripping canoeists, but will not matter at all to the cottager who wants only an occasional paddle down the lake.

- *Write a good outline first.* Outlining requires grouping similar items together into categories that can be identified with clear, descriptive headings. It also requires that the categories be listed in logical sequences and divided into logical subdivisions. This is the essence of using classification and division to organize your report. A good outline is an information hierarchy with consistent, balanced categories. It includes all the categories that describe the main topic, leaving out nothing of significance. However, writing one requires knowing all your research information in detail, having good judgment about which data sets belong together, and being able to express the organizing characteristic for each category in a simple topic heading. We have already explained the benefits of this method of organizing with an outline: you gain ease of drafting the report and quality in the final product, and your reader gains ease of understanding your information and accessibility to the specific topics that he or she needs to read.

- *Use only one characteristic at a time to classify your data.* For example, if you wish to classify the coins in your collection, you may select denomination as the first organizing characteristic. However, after you have gathered all the $2, $1, and $0.25 pieces into piles, you may then sort the remaining coins into silver, copper, and alloy coin categories. The change in organizing characteristics makes the whole classification system invalid and confusing—not useful when you are attempting to identify and access individual pieces in the collection.

Cause and Effect

The *cause-and-effect* pattern works in two reporting situations. The first is the report discussion that seeks to predict the outcome or effect of a given action or set of actions, based on the causes: "If we raise the voltage to the oscillator, what effect will it have on the intermediate-amplifier output?" "If the government passes regulations against using polyethylene liners in landfill sites, what alternative materials are possible to seal leachate from the water table?"

The second is the report discussion that describes an unpredicted phenomenon, or effect, and seeks possible causes: "When we installed bluebird boxes in this area, why did the squirrel population increase?" "When we added reinforcement to this end of the structure, why did the other end shift slightly?"

Examining cause-and-effect relationships in science and technology can be problematic. Environmental factors are often complex and vague, difficult to identify and isolate experimentally. It is therefore important to describe such causal connections carefully in your report, with attention to as broad a range of detail as possible. It is also important to reason through your information logically, using conventional and accepted forms of analysis.

Use the pattern of topics for the research report found in Chapter 4.

- Describe first your research goals, along with the reasons for them if relevant, and the methodology you used to investigate causes and effects.

- Describe your results in terms of either cause-and-effect or effect-and-cause: "The following are the alternatives to polyethylene liners that our team considered viable," or "The following are the test results when the team investigated the structure after it shifted." It is not always easy to know when you have all the facts. Your own training, experience, and common sense are usually the best guide. When you are satisfied that you have considered all factors, it is time to write the report.

- Analyze the data in a discussion section. Share your reasoning with your reader. How you add up the factual information will determine how persuasive your argument is. Examine your assumptions carefully and discuss them with your reader. Show the analytic techniques you used. Make sure they are appropriate to your audience. For example, detailed mathematical analysis of data will not be useful to a manager who wants to read only your conclusions and recommendations.

- If required by the situation, end with a section of recommendations. Like the choice of data to present and the methods of analysis, your choice of recommendations will depend on your own training, experience, and common sense.

Errors in Logic

Errors in thinking can creep into your scientific and technical work. Always include a double-check step in your working processes. The beauty of writing about work is the opportunity it gives to think about your work and see it again objectively, to ask yourself whether errors were made in collecting and analyzing data or in developing solutions.

Errors in a scientific or technical discussion of causes can be one of two kinds. Assuming a causal relationship where none exists is one kind of error; assuming no causal relationship where in fact one exists is the other kind of error that occurs when using the cause-and-effect pattern of organization. Writing allows you to do a double check of your assumptions and catch such errors.

A scientific or technical report uses the language of statistical probabilities. An investigation rarely proves a case beyond any doubt. It can only reduce the probability of error to an acceptable level and increase confidence in test results.

Thinking about causes and effects is a large undertaking. A practical approach in your writing is to revise your report drafts looking for specific errors in logical thinking that technical writers often make in discussions of cause and effect.

Appeal to the Person (Ad Hominem)

Avoid personalities and personal bias. In Latin, *ad hominem* means "to the man or person." If your analysis focuses on the character of a person or group, not the quality of their reasoning, you are diverting attention from the facts and analysis. Your report should have an *ad rem* (Latin: "to the thing") focus, emphasizing the evidence.

This error can occur in your writing when you examine evidence or discuss alternatives that you do not like. Personal bias can appear as name-calling, prejudice against various groups in society, or associating the alternative with unpopular social groups.

Circular Reasoning (Begging the Question)

Don't assume the truth of your statement—instead, you must prove it. "The Canon printer is the best laser printer because it is better than the others." A statement like this simply re-states in the second part of the sentence the assertion in the first part. We have used many different words to say the same idea. Any such statement adds nothing to the argument and does not prove the initial assertion to be true.

Watch for any statements in your report that begin with "X is good because it [some specific example of good]," where "good" can take many forms: "The Pentium microchip is valuable because it processes data faster."

Sampling Errors

Many environmental phenomena can be causes or effects, or can influence a cause-and-effect relationship between other factors. Therefore, to carry out your investigations it is essential to gather as much information as possible under the constraints of time and resources. Choosing a sample large enough and composed correctly enough to draw reliable conclusions is a matter of expertise in statistical methods.

We are all guilty of drawing general conclusions based on fragmentary evidence. "The Renault is an unreliable car because my friend's Renault is unreliable." This is sometimes called "jumping to conclusions," or, as someone once remarked, "extrapolating on the basis of one sample."

Error of "Post Hoc Ergo Propter Hoc"

In Latin, "post hoc ergo propter hoc" means "after the thing, therefore because of the thing." In this error of logic, we assume that the second of two observed events caused the first. In truth, if Event A comes before Event B, it doesn't necessarily cause Event B, although there may be a scientific correlation. In your own lab and fieldwork, you observe many events during the course of testing. When doing causal analysis of the results, look for causes in the order of events, but look beyond to the specific mechanisms that prove a causal relationship.

In order to prove that one event or circumstance caused another, you must show that the first event happened before the second. You must also prove a cause-and-effect relationship between the two events. The mistake in logic occurs when you prove that Event A came before Event B, and then just assume that A caused B without showing specifically how that occurred: "Bluebird boxes cause squirrel populations to increase because we counted more squirrels after we put up the boxes."

Problem/Solution

The *problem/solution* pattern of report organization comes from the reporting situation itself. Your boss has encountered a problem that affects your department or company. He or she then asks you to investigate the problem and report on it, including options for solving the problem. This pattern of organization follows the natural pattern of work. The sequence of sections in your report is the same as the sequence of steps that you took to confirm, investigate, and analyze the problem, finishing with descriptions of possible solutions.

This pattern also applies to business opportunities. Your boss may not have a "problem" as such, but he or she may see an opportunity in changing current engineering or technical practices, or see a chance at developing a new product. Improving current systems is an ongoing process in business and industry. The sequence of work and of sections in your report will be the problems/solutions pattern.

Use the following pattern of topics for your report.

- Describe the circumstances of the report: the problem and your mandate to investigate it, or the opportunity to improve existing products or systems.

- Confirm the problem. Your initial investigation will be an environmental scan, looking at the circumstances surrounding this problem or opportunity and determining whether or not a problem or opportunity exists, and if so what its potential impact on your company might be.

- Describe the methods you used to investigate the problem.

- Describe the results of your investigation. Use tables, graphs, or other visuals where appropriate to show this information. Summarize your results in a vertical list organized from most to least important results.

- Describe options for solving the problem or taking advantage of the opportunity. List them with the most important option, from your viewpoint, first.

- Write a separate section at the end indicating which option you recommend (if the reporting situation requires you to make recommendations). Include a review of the data and the reasoning you used to make the choice.

When investigating and reporting on a problem, remember that your reader, usually your immediate supervisor, will want to make his or her own choice of actions in the matter. To meet this need, set out clear, specific, well-organized information regarding the nature of the problem, the methods used to investigate it, the results of your investigation, and the likely options for solving the problem. Your own training and experience are again the best guides as to how to investigate the problem and which data to report.

Scientific and technical researchers know that your choice of research methods will determine the kind of information that you develop. Therefore, it is important in the problems/solutions report to describe your reasons for choosing a particular method or set of methods for investigating the problem. This permits better analysis of the results and boosts the reader's confidence in workable solutions when you describe them.

Do not overstate the solution. It is natural, as you become involved in the investigation, to develop a sense of which option is the best solution to the problem. It is important to be aware of your own biases in this regard. This natural tendency can result in promoting one solution over another in various ways, or in promising results from one course of action beyond what is reasonable or even possible. Check your report drafts when you revise and edit for clues that you have overstated the solution; that is, where you have made statements that tend to emotional language, or have included sections that describe some solutions in greater detail than the others.

Write an overview in your introduction: your reader will need a summary of the problem, the methodology, and the proposed solutions. Format your work with descriptive headings and provide the table of contents and illustrations elements that will guide your reader to specific sections of your text.

PLANNING YOUR REPORT

Report Elements and Report Graphics

As you develop an outline for the content of your report, make two important lists: the report elements you will need to meet your reader's format and access expectations, and the report graphics you will need to support your text descriptions and provide information in standard formats.

Report elements are the pages of front matter before the text, including the title page and table of contents, and the back matter after the text, including the reference list and the appendix. They provide key information about the report, summarizing and locating information within the text and other report elements. Plan your report elements based on your reader's needs and expectations. Then, as you write the first drafts of the text, you can include information from the text required by report elements, such as page numbers required by a table of contents or reference entries required by citations of secondary sources in the text.

Plan your report graphics with a preliminary list. As you read your notes, look for field or lab data that must be summarized in your report as a table, graph, or chart, because these are conventional methods of reporting data, because they are a compact form of presentation, and because they are easier for your reader to interpret and understand.

Also look for diagrams, charts, maps, or photos in your research materials that provide clear, convenient explanations of scientific or technical processes or instrumentation. Scientific and technological structures, theories, and processes must be visualized to be understood clearly. A good technical report balances good writing with good graphic communication.

CHECKLIST FOR REPORT RESEARCH AND ORGANIZATION

Be sure that you have ...

- Identified the primary audience, type of reader, any secondary audience
- Identified the purpose, and stated it clearly in one or two sentences
- Understood your role on the team, attended regular team meetings, identified team strengths and weaknesses, avoided pitfalls of team writing, and engaged in effective team communication
- Gathered all primary and secondary research materials
- Made summary notes of key points; recorded information about secondary sources
- Written an outline; organized all information into a hierarchy of headings, clear topic headings, correct subordination, and logical sequence of topics
- Checked your outline for patterns of organization
- Checked your outline for errors of logic
- Checked your outline for format errors
- Checked your outline for faulty co-ordination, subordination, or parallelism
- Made a list of required report elements and report graphics

Chapter 2

The Report-Writing Process

Overview: This chapter looks at how you actually write a report—the process of setting your thoughts and ideas into words and pictures—and discusses the principal activities in writing: drafting, revising, and editing.

STEPS IN WRITING A REPORT

Before you look in detail at ways to write a successful report, read the following overview to help you plan effective use of your writing time.

1. *Gather all available research materials.* Primary sources may include your field or lab testing results, notes and journals, and e-mails or memos sent during a project. Secondary sources may include print materials from libraries, electronic files, and company documents.

2. *Plan before you write.* Use a formal outline to create a logical, hierarchical sequence of descriptive headings for your report, based on your research notes. This plan will speed your work of writing, keep your report organized, and provide your reader with descriptive topic headings needed for easy access to your information and ideas.

3. *Start drafting with the main points of the body.* Write down your headings and paragraphs of description under each one. Use your research notes to develop complete, accurate descriptions of the material, focussing on key words and phrases. Use your outline to get them in the right order under logical, descriptive headings.

4. *After the main text, write your concluding sections.* These will include a summary of the most important points and an analysis or reflection on them. If your audience and purpose require it, add a section of recommendations.

5. *Add your graphic presentations.* Choose tables and figures to add essential information and visualization to your main text. Illustrations are essential to good scientific and technical reporting. Design them to communicate complete, accurate information and to support comments and inferences drawn in the text. Integrate them fully with your text by including appropriate numbers, titles, labels, and textual references. Include graphics not essential to your discussion as appendices.

6. *Write your introduction.* The introduction sets the reader's expectations in the form of the report's plan and purpose. Describe the logical order of headings in your report and the reasons for writing it. Include all the information your reader needs to understand the content of the report, such as a theory of operation or a map of the field study area.

7. *Finally, compile the report elements.* Be sure you have the ones your reader needs. For example, if you have presented lab test results summarized as tables, you will need a list of illustrations. Assemble the information for each element and write it in the conventional format. As you place the elements into the report, check that you have the correct information in each element, the correct format for each element, and the elements in the correct order.

CREATING YOUR TEXT

Create the text of your report in three steps: drafting, revising, and editing.

1. *Drafting* means translating your research notes into words, sentences, and paragraphs to express the essential content. Many student writers think the text is complete after the first draft. Professional writers know that this is only the beginning.

2. *Revising* is the process of altering the content of your report. Adding, omitting, and re-ordering your content is the most important step in making contact with your reader and communicating the right information. Revising ensures that all the material useful to your reader is covered and that all appropriate descriptions of scientific and technical principles and practices are included.

3. *Editing* is the process of correcting your text: its style, its expression, its grammar, and its spelling. Correct English is universal English, understood by everyone with a fundamental grasp of the language. Precise technique is essential in a science lab to produce accurate, reliable test results; precise English is essential in a report to produce a document that communicates data and concepts accurately and reliably to the widest possible audience.

DRAFTING

As you translate your research notes into words, focus on straightforward, clear descriptions of field or lab procedures and technical instrumentation. Professors and professionals working in your field appreciate being able to read about your work in simple, clear, descriptive terms. Your descriptions of current techniques and innovations will be essential to their business.

However, few people can create clear, simple texts the first time they express themselves in written words. Most writers find it difficult to choose among many possible ways to express an idea. Many writers are afraid to fail, afraid to find that their own written descriptions are confusing and full of mistakes, and afraid to re-write a report section several times, trying out different words and phrases to capture a difficult concept. Many writers worry about spelling, grammar, and usage at this early writing stage, instead of focussing on essential information and leaving matters of spelling, grammar, and usage to the end of the writing process.

Starting to Write

Start with the easy ideas and information. For primary research, describe your methods and results. Use the key words in your notes that express important data and essential ideas, and weave them into your own sentences and paragraphs. For secondary research, organize your information in the order of your outline headings.

Focus on your outline's sequence and your research notes in order to create clear, direct descriptions. Write each heading and read the research material that applies to that heading. Then write your own version of that information, using the key words and phrases from your research notes to express the central concepts. Build your sentences around them, especially topic sentences in paragraphs. Keep each paragraph short and simple; break complex processes into small steps, and describe each of them in a single, short paragraph with a strong topic sentence. Use bulleted or numbered lists for all lists of more than three items: steps in procedures, equipment or materials needed, and so on.

Draft quickly; do not pause to rewrite sections of your text or to correct mistakes. Look at your notes, not at the words appearing on the computer monitor. Trust the writing process—if a section proves hard to write, skip it and go on. You can return later with fresh expressions of the essential ideas.

Your first draft will give your document structure, but not its final form. Do not try to make it perfect the first time. Very few professional writers can produce a near-perfect draft the first time, considering all aspects of the text at once. The most effective method of writing is to leave revising and editing to later drafts. Allocate time in your work schedule to revise and edit your text so that it will be accurate and correct.

Writers take regular breaks from the job of writing. However, if you stop at the end of a section, you will have to think through the issues of the next section of text in order to get started again. Stop in the middle of a paragraph or section so you can start writing again more easily next time because you will know immediately how to end your paragraph or section. Ideas will flow.

Try to type with your eyes on your notes, not on the computer monitor. Ignore typing mistakes on your display and focus on your information; your content will be better, and you will be using time more efficiently.

Using a Word Processor

Most professionals write using computers with one or more word-processing applications and access to a printer. These new technologies affect our writing, so learn how to take advantage of technology's power and avoid some of its pitfalls. The computer can act as a kind of alter ego, entertaining as it helps you write. An example is the Microsoft Word™ Office Assistant, which you can summon at any time to answer questions about word processing—the Assistant is usually accompanied by amusing sounds and actions, and you can customize its form to suit your whim.

The speed and power of word-processing software benefit the scientific and technical writer in many ways, but they also impose burdens. Writing with a computer is different from writing with a typewriter. Computers can

- process information quickly and accurately
- highlight errors in your text
- communicate with the outside world, accessing from a variety of sources (CD-ROM, the Internet, e-mail, scanners, digital cameras) information that is vital to the success of your report
- store your developing text files
- place scientific and technical data in the form of tables and figures on the same page as text, so your reader can get the full picture
- print clean, fully illustrated pages, similar to those previously possible only with offset typesetting.

However, power also comes at a cost. Ease of use can lead an unwary writer into lazy writing and editing habits—the speed of the computer can cause you to draft, revise, and edit too hastily, without taking the necessary time to reflect on content, style, and format. The computer, with its many abilities for display and communication, can shift a writer's attention away from the content of documents and important ways to connect with the reader, such as providing illuminating examples or adding a map to clarify spatial relationships. In addition, the variety of features available in word-processing software requires a writer to spend considerable time to learn effective ways of using it.

Keyboarding skills give you speed and accuracy, as well as a sense of comfort with the computer. However, if you hunt and peck at the keyboard as you compose, that is fine too—when putting words together to express yourself, slowness can be a virtue.

Writing Scientific and Technical Descriptions

Most of your writing will be descriptive. It will describe objectively and in detail the steps taken in field and lab work and the measured observations of results.

Learn to look for exact quantities and measurements, and write them into your text. "The emitter current on the power transistor was very large," tells the professional and technical reader little useful information. "The emitter current on the IC-12 NPN-type power transistor peaked at 250 mA when a bias voltage of 5 mV was applied to the collector," tells your reader a great deal more that is useful.

Keep your paragraphs short. A single paragraph covering an entire page of report text is much too long. Write short paragraphs about smaller, more specific aspects of each topic. Introduce each paragraph with a strong topic sentence that clearly states the type of information you describe in that paragraph.

Using Secondary Sources

The best use of your research material is paraphrasing, or putting the facts and ideas into your own language. Read your sources carefully and understand the content before trying to write it down. Do not attempt to substitute key technical and scientific terms such as "photosynthesis" or "motherboard." They have no simpler equivalents that mean the same. Instead, make sure your sentences containing key words are substantially different from the original and focus on your own audience and purpose.

Use direct quotations sparingly, if at all. In most cases, they should not be longer than 75 words. All direct quotations must be set apart in the report text with quotation marks. Quoted passages longer than 75 words must also be set in a separate paragraph, single-spaced, with margins increased to 5 cm (2 in.).

Always cite your sources as you write. At the end of a paraphrased or quoted section, insert a citation in the format appropriate to your field. If you want to revise or move the paragraph or section later, simply move the citation with the text.

Most students know that they should cite direct quotations, but many miss the requirement to cite all paraphrased or graphic material. You must cite all borrowed material, no matter what form it takes in your report.

REVISING

When you have finished the first draft, put your text away for a period to gain perspective on what communicates and what does not. Be sure to leave enough time before your report deadline to do this. If you practise revising on your computer monitor you will avoid printing several drafts of your unfinished report, saving you time and paper.

Revising Your Text

Revise the content of your report first. Check the sequence of main points to ensure they are complete, in the correct order, and get all the important detail into the report. Look again at secondary detail. With the full text written, you will have a better perspective on what to include for your audience. Ask yourself if explanations of scientific and technical methods are too lengthy, or lack detail. Revisions of content have the greatest impact on your reader. They will significantly improve the organization and readability of your report.

Revise the style of your report next. Readers tend to focus on and remember specific details of general principles, so consider adding some. For example, "Mickey Mouse demonstrates the principle of *neoteny*, the retention of juvenile characteristics into adulthood."

Check to ensure adequate transitions: words, sentences, and paragraphs that signal a change in topic or clarify relationships between ideas. Read over your descriptions for the shortest possible expression. We have a tendency to talk around ideas when unsure how to say them. This creates more words than necessary in a first draft. Written work should be concise: the right number of words to express the idea, no more and no less. Cut wordy language in favour of short, direct language. For instance, use "because" instead of "due to the fact that."

Using Good Scientific and Technical Writing Style

Keep in mind your writing goals as you review your drafts. The following qualities are characteristic of all good scientific and technical writing:

- *Accuracy*. Record facts carefully in lab work, in the field, and when researching the literature. Read and record your sources accurately. Be objective and free of bias. Scientific and technical writing must be reasonable, fair, and honest.

- *Accessibility*. Make it easy for your reader to locate information in your writing. Write in small, independent sections. Use an outline of headings showing main topics divided logically into smaller units. Use the elements of your report to show your reader where to find these topics.

- *Comprehensiveness.* Provide all the information your reader will need: background, methods, principal findings, conclusions, and recommendations. The detail you describe must be effective, efficient, and safe. The report is also a record of projects, meetings, and events that a company or agency needs in its planning process.

- *Clarity.* Your writing should convey a single meaning that readers can easily understand. Watch out for ambiguous words—ambiguity is good in creative writing, bad in scientific and technical writing. Unclear technical writing is expensive, breaks down co-operation, and can be dangerous.

- *Correctness.* A *convention* is the way people usually do things. Language has conventions: grammar, spelling, punctuation, and usage. Documents have format conventions, such as letters, memos, and reports. Learn the conventions and apply them. Readers of scientific and technical literature around the world use many different kinds of English. To communicate with them, it is essential to write in standard, conventional English.

Avoiding the Passive Voice

Some scientific writers will tell you to use the passive voice in every sentence of your report: "It was observed that the male lionfish was inactive when it was fed during daylight hours." While this technique was acceptable in the past, mainly to project the objectivity of the scientist, it is not the current style. Using the passive voice throughout your text leads to loss of clarity, awkward phrasing, and needlessly long and complex sentences.

The modern style is to combine the third-person, active voice appropriately with the passive voice to focus the reader's attention on the scientific and technical principles and processes discussed in the report: "The logs float into the catch basin and are hauled into the mill by a chain conveyor." This example focuses attention on the logs at the centre of the milling process.

Creating a Formal Tone

Bear the following tips in mind when revising:

- Use the past tense, except when describing universal processes, as in the following: "Changes in water temperature reduce resistance in the sensor and increase current flow in the meter circuit." The tone should be formal, not personal and informal, so avoid conversational language: "Hey, man, like, check the ammeter level before you close the case."
- Avoid contractions (don't, they'll, I've)
- Use the third person: "The investigations included a control plot of mixed conifers planted at intermediate spacings." Avoid the first and second persons, such as: "We included a control plot of mixed conifers planted at intermediate spacings in our investigation," and "You must include a control plot of mixed conifers planted at intermediate spacings in your investigation."

EDITING

Editing is correcting. Writing in correct, Standard English communicates to a wide audience, including those for whom English is not a first language. Mistakes in style, grammar, and usage come from a variety of sources. Be aware of the mistakes you typically make and check your report draft carefully for them.

Using Correct Language

Correct language is clear language. Violations of Standard English will muddy your meaning and suggest ignorance. If you write "I seen" and "I done," or confuse the verbs "lie" and "lay," or say "myself" when you mean "I" or "me," your audience will assume you know no better. Worse, your audience will begin to focus attention on your mistakes and miss what you have to say.

Correct language is learned in steps. First, learn the parts of speech in a sentence: verbs, nouns, adjectives, adverbs, prepositions, conjunctions, and so on. Second, learn the uses of words and groups of words: subject, predicate, object, phrase, clause, participle, gerund, and so on. Third, remember that most errors are common (comma splices, sentence fragments); learn the ones you tend to make, and watch for them.

Correct language is a responsibility of all report writers. Do not think of correctness as a "basic" skill; writing clear descriptions in a report does not come naturally. It is the product of hard work and attention to detail, qualities that also make a good scientist or technician.

Correct language is consistent. When you make choices of style, technical terms, spelling, and so on, make them consciously and consistently from one paragraph through to another. Inconsistencies create confusion and misunderstanding, so edit for consistency. Help with language problems is readily available: Internet sites, campus libraries, or any good bookstore will have resources to assist you with any specific problem.

The following is a brief list and explanation of some common errors found in scientific and technical writing.

Avoiding Errors of Grammar

Grammar is the structure of an English sentence—called "syntax" by experts in linguistics—and the forms of words. Especially in English, each sentence depends for meaning on the *order* of words and on the *form* of words, such as the different forms of the verb *to drink*: *drink*, *drank*, and *drunk*.

Most writers have some sense of English grammar. This comes from automatic language learning when we are very young. Such learning, a function of the human brain, unfortunately includes errors. At school, we learn about language and develop an understanding of syntax; unfortunately, this understanding is incomplete and includes errors. Most writers say, "I know what sounds right." This flawed test of grammar is simply applying what we have learned long ago, including the errors. Fortunately, some errors of syntax are common to most writers. Learn them and look for them in the final drafts of all your documents.

Sentence Fragment

A *sentence fragment* is an incomplete sentence: "Determining binding constants." An English sentence must have a subject (the person or thing discussed in the sentence) and a predicate (the action or state of being of the subject). A sentence missing one or both of these elements is a fragment. It seems much easier to spot fragments in someone else's work than in your own; enlist the help of a friend or colleague in reading your report draft for errors.

Run-on Sentence

Many writers erroneously believe that a run-on sentence is simply too long; however, a sentence containing too many main ideas or too many modifiers is also an error of style. The error of grammar called the *run-on sentence* is an error of sentence compounding. A compound sentence contains two main ideas, or principal clauses: "Dogs bark, and cats meow." We usually join principal clauses with a co-ordinate conjunction: "and," "but," "or," "so," and "yet." However, we can also join principal clauses with a semicolon when the ideas they express are closely related: "Dogs bark; cats meow."

In the *comma splice* form of run-on sentence, the writer has joined principal clauses with a comma: "Dogs bark, cats meow." The comma is only a grammatical pause, marking divisions between subordinate units. Do not use a comma to join principal clauses.

In the *fused* form of run-on sentence, the writer has joined principal clauses with nothing: "Dogs bark cats meow." This makes it difficult for the reader to sort out main ideas within the sentence.

Agreement of Subject and Predicate

Subjects are nouns or pronouns; *predicates* are verbs or verb phrases. Both nouns and verbs have singular and plural forms:

An external syringe pump circulates the sample.

External syringe pumps circulate the sample.

Note that we add "s" to the noun to make it plural; we add "s" to the verb to make it singular. Know the correct singular and plural forms of nouns and verbs, or check them if uncertain. The subject must agree with its predicate, singular or plural. In the first example above, the singular "pump" agrees with the singular "circulates."

Two factors make this error hard to spot for many writers: long modifying phrases or clauses between the subject and predicate, and sentence subjects that create uncertainty about whether they are singular or plural. Consider the following sentence with phrase modifiers:

The high stability of such instruments, combined with reference surfaces for detecting nonbinding constants, permit refractive index changes to be measured and account for sample or instrument temperature drifts.

If we extract the bare subject and predicate from the phrase modifiers, we can see the error clearly: "The stability permit changes and account for drifts." The singular subject needs a singular predicate: "The stability permits changes and accounts for drifts." With its modifiers added back in, the correct sentence reads like this:

> The high stability of such instruments, combined with reference surfaces for detecting nonbinding constants, permits refractive index changes to be measured and accounts for sample or instrument temperature drifts.

Consider the subject of the following sentence:

> Every technician and lead hand know the recommended safe procedure.

The words "every" and "each" (and their compounds "everyone," "everybody," and so on) are singular in meaning: one of many. Thus, the predicate that matches must be singular:

> Every technician and lead hand knows the recommended safe procedure.

Agreement of Pronoun and Antecedent

Every pronoun has a noun preceding it somewhere in the text to which the pronoun refers. That noun is called the *antecedent* of the pronoun. The pronoun must agree with its antecedent in gender, number, and case. Consider the following sentence:

> John knew that he had completed the procedure successfully.

The pronoun "he" agrees with the antecedent "John" in gender (both are masculine), number (both are plural), and case; "he" is the subject form of the pronoun required by its use as subject of the subordinate clause "that he had completed the procedure successfully."

Some antecedents refer to people but, unlike "John," without gender. Consider the following example:

> Everyone carried their own field pack.

The antecedent of the pronoun "their" is "everyone," which is singular in meaning. "Their" is plural and does not agree. The singular pronouns that agree are "his" and "her." So, use both; modern style demands inclusive language:

> Everyone carried his or her own field pack.

Errors in the Principal Parts of Verbs

Every English verb has three principal parts: the simple present, the simple past, and the past participle. The dictionary lists the simple present form. If no other form is listed, the verb is regular and adds "-ed" to form the simple past and past participle. If the verb is irregular, the other forms of the verb will be given: *drink*, *drank*, and *drunk*.

Most errors occur in the use of irregular verbs. If you suspect an error in your writing, check verbs forms in a dictionary. If an error is present, note the correct forms and use them. The following is a list of commonly misused irregular verbs:

Simple Present	Simple Past	Past Participle
drink	drank	drunk
swing	swung	swung
break	broke	broken
freeze	froze	frozen
begin	began	begun
set	set	set
lie	lay	lain
lay	laid	laid

Note the difference between "lie" and "lay." To lie is to place your body in a horizontal position (we are not concerned here with the regular verb "lie," meaning to tell an untruth); it is an intransitive verb followed by a predicate noun or adjective: "He has lain asleep for two hours." To lay is to place or put; it is a transitive verb usually followed by a direct object: "I laid my textbook on the computer desk when I had finished."

Misplaced Modifier

Place modifying phrases and clauses close to the words they modify. The logic of English is order. A reader will associate a modifying element with the words closest to it. If these words are not the intended association misunderstandings will arise—sometimes in a comical way:

The technician reported that the construction was completed in his e-mail.

The modifier is "in his e-mail." The logical association is with "reported." By leaving the modifier at the end of the sentence, the writer associates it with "completed." Readers can eventually find the correct association by rereading the sentence, but this takes time and builds annoyance. Correct the error in your drafts:

The technician reported in his e-mail that the construction was completed.

Dangling Modifier

Check that your modifiers have a word to modify, usually a noun or a verb. We sometimes lose the intended associations with modifying phrases and clauses in the process of composing sentences.

Having completed the assembly, the instrument calibrated to within 10 percent of its normal range of values.

The modifying phrase "having completed the assembly" will associate itself in the reader's mind with "the instrument" because it is the closest noun in the sentence. The idea that the instrument completed its assembly by itself is comical. Restore the correct association by adding the noun that belongs with the modifier:

Having completed the assembly, the technician calibrated the instrument to within 10 percent of its normal range of values.

Punctuation

Place punctuation marks to identify grammatical units. If you are not sure how to identify grammatical units in a sentence, or which punctuation marks are correct, do not punctuate your sentences with inappropriate or incorrect punctuation marks. Mistakes in punctuation indicate an ignorance of grammar and can seriously mislead a reader. Punctuation can completely change the meaning of a sentence:

The staff says the boss is incompetent.

The staff, says the boss, is incompetent.

The most commonly used punctuation marks are discussed below.

Comma

The comma separates grammatical units in a sentence and is a kind of grammatical pause. Good scientific and technical writing uses short sentences and paragraphs that require fewer commas, but the comma is still the most commonly used—and misused—punctuation mark.

In *compound sentences*, place a comma at the end of each principal clause:

> The co-efficient of friction rose as heat was applied, but thermocouple output remained constant.

In *series*, separate each pair of items in a list or series with commas:

> Follow instructions carefully when unpacking the instrument, assembling it, calibrating it for water temperature and dissolved ions, and mounting it in the personal carrier.

To separate *introductory phrases and clauses*, place a comma after introductory phrases and clauses before the subject of the sentence:

> After assembling and mounting the instrument, uncoil the cathode lead and trail it in the watercourse behind the unit.

Use commas to set off *nonrestrictive elements*. Some modifiers are essential to show the reader the complete meaning of the words that they modify. Other modifiers are less so; they add information about the words they modify, but they are not essential to the meaning of the whole sentence. We call essential modifying elements *restrictive* because they restrict the meaning of the words they modify to a specific group. Consider the following:

> The college that Mary attends has a linear accelerator.

The restrictive element here is a clause: "that Mary attends." It restricts the word it modifies, "college," to just the one she attends. If we drop the clause, the sentence reads "The college has a linear accelerator," leaving us to wonder which college is meant. Enclose nonrestrictive elements in commas. Note that you must use two commas; it is a common error to place only one. Consider the following example:

> The efficiency of any drill bit, whether used on a cable tool or rotary drill, is compromised by the incorrect use of drilling mud.

The nonrestrictive modifier, "whether used on a cable tool or rotary drill," can be omitted without compromising the sense of the sentence.

Note that *appositives* are short, nonrestrictive elements:

Gregory Dudek, a professor of robotics at McGill's Centre for Intelligent Machines, has developed processes for localizing a robot with minimum travel.

With *places and dates*, the use of the comma is conventional, standard practice:

The neutron probe for measuring total moisture content in pavement structures was developed in Airdrie, Alberta, a suburb of Calgary.

The Confederation Bridge linking Prince Edward Island to the mainland was completed on June 1, 1997, and will be turned over to the Government of Canada in 2032.

Semicolon

Use the semicolon instead of a co-ordinate conjunction to join principal clauses that are closely related in meaning:

This analog test signal is not passed through the smoothing filter of the direct analog converter; instead, the output of the oscillator is connected directly to the multiplexer at the analog-drive converter input.

Note the use in this example of the conjunctive adverb "instead" to link the ideas in the two principal clauses. Use the semicolon also in some series; when any item in a series contains commas, use the semicolon instead of the comma to separate items in the series:

Do not attempt bioengineering solutions in any of the following situations: severe soil, air, or water contamination; degraded stream bottom; uncontrolled human or animal traffic at the site; or too much shade for selected plant species to thrive.

Colon

Use a colon to introduce a list of additional facts after a complete sentence:

The Civil Engineering Research Foundation (CERF) identified the following research goals in the architectural industry, in order of importance: computerization, the improvement of design technology and practice, the implementation of computer-integrated design and construction, and the development of new flexible, multi-user design tools.

The colon is a full stop, like the period. Do not place a colon in front of just any list in the sentence, like the error in this example:

SAW devices include: linear resonator and resonator-filter devices, linear devices using unidirectional IDT's, linear devices using bidirectional IDT's, and nonlinear devices.

In this example, the words after "include" are collectively the direct objects of the verb, a natural grammatical relationship. No punctuation is required at all.

Avoiding Errors of Style

Scientific and technical writing describes and explains complex processes, products, and principles, so keep the structure of sentences very simple—single principal clauses with simple subjects and predicates. The modifying elements in the form of scientific and technical terms will make the sentence longer and more elaborate. Adding complex syntactical structures to complex terminology will obscure meaning even for the most knowledgeable reader. The following are specific ways of keeping your style simple.

Ambiguous Pronoun References

Pronouns like "it" or "them" must refer to a specific noun used previously in a sentence or paragraph. This noun is the pronoun's antecedent, and your reader will instinctively look for it. If the correct antecedent is not close to its pronoun, your reader will become confused between possible antecedents or associate the pronoun with the wrong antecedent.

After mounting the microbial sensor in the metalworking fluid system, connect the recorder and monitor it.

Does "it" refer to the recorder, the metalworking fluid system, or the microbial sensor? Your reader will not know with any certainty. Restructure the sentence to make your meaning clear, as in the following example:

1. Mount the microbial sensor in the metalworking fluid system.
2. Connect the recorder.
3. Monitor the recorder output from the sensor for ten minutes before leaving the installation to ensure the accuracy and stability of both sensor and recorder.
4. Check the recorder every third day, and download its memory to a PDA.

Diction

Choosing the right vocabulary, or *diction*, is essential to good scientific and technical writing because of the specialized and often complex terms needed to describe and explain scientific and technical processes and principles. Use common words; not "He inaugurated the investigation," but "He began the investigation."

Parallelism

Make sure lists in your sentences all contain the same grammatical units. *Compounding*, or *co-ordination*, is one of two basic structures in an English sentence. The other is *subordination*. Compounding is joining two or more elements to a sentence using a co-ordinate conjunction ("and," "but," and "or," along with variations "nor," "either–or," and "neither–nor"). We may also include their American cousins "so" and "yet."

The equipment packed for each unit fire crew includes pumps, hoses, stranglers, spare nozzles, and portable radio transceivers.

Write long compounds in the form of a vertical list, usually with bullets:

The equipment packed for each unit fire crew includes the following:

- 2 Wajax pumps, one primary and one backup
- six 100-ft sections of #4 hose
- 2 stranglers
- 2 nozzles, one primary and one spare
- 2 portable radio transceivers

Note that each item in the bulleted list is a noun with its modifiers. This list is a parallel construction. When list items are not parallel, meaning becomes unclear:

This test for desaturase enzyme in plants requires small quantities of plant material, completed in about ten minutes, simple lab equipment and chemicals, and accurate results.

What follows the subject of this sentence is a list that includes verbs, adjectives, and nouns, all different grammatical units. To make the list parallel, make them all the same units, for example, verbs:

This test for desaturase enzyme in plants requires small quantities of plant material, can be completed in about ten minutes, needs only simple lab equipment and chemicals, and produces accurate results.

Avoiding Errors of Usage

Writers frequently misuse words in ways that fall outside the categories of grammar and spelling errors. Check meanings frequently in a good dictionary and by using words in your own sentences. The following are some common errors of usage.

Accept, except. "Accept" means to receive:

The jack for the communications port accepts a standard, 9-pin flat plug.

But "except" means to exclude:

The external surface of the robot arm is single-mesh steel except for the alignment sensor mounted on the end to aid its manipulation.

Alot. You wouldn't say "alittle"; don't write "alot." Write "a lot," meaning "many," or use the more formal "a great deal" or "often." The best practice for scientific and technical writers is to omit such vague expressions entirely. Quantify your expressions precisely: "3.5 grams," "512 forest fires," "600 microns," and so on.

Amount, number. This error is now so common that many readers ignore it. However, many do not. "Amount" refers to things you cannot count individually; therefore, you must measure their quantity by weight or volume, like wheat or gasoline. The words associated with amounts are "little–less–least" and "much–more–most":

The amount of medical imaging in Canada will increase as the technology becomes a more digital, distributed process.

"Number" refers to things that you can count individually, like pennies or books:

The number of live cuttings that can be placed in rip rap to improve soil stability depends on rock spacing and the slope of the bank.

CD-ROM disc. CD-ROM stands for "compact disc read-only memory." The word "disc" is redundant. The same is true for DVD disc, since DVD means "digital video disc."

Data. The word "data" is the plural of "datum." Like "criterion–criteria" and "medium–media," it has Greek origins and does not follow regular English plural forms. Some confusion exists among scientific and technical writers, journalists, and editors regarding whether "data" is plural or singular. The safest choice is plural:

The projected data for atmospheric carbon loadings by 2032 are listed in Table 2.

Effect, affect. Do not confuse these. As verbs, "effect" means "to bring about a change" or "to accomplish," while "affect" means "to influence":

The addition of solvents to the solution effected the desired change in the gram-molecular weight.

but:

Variations in sample temperature did not affect the accuracy of the study's flow measurements.

As nouns, "affect" is a feeling or emotion; "effect" is a result, a change produced by an action:

The laser beam in a turbid medium had the effect of scattering photons into ballistic components.

If, whether. "If" introduces a condition:

VRML authoring tools have to become more sophisticated if designers are expected to take them as serious development media.

"Whether" introduces an alternative, a choice of two options, usually the choice of doing something or not doing it:

Designers have discussed whether to expand the Indexed Line Set to allow VRML models to display different line thicknesses, patterns, or types.

Many writers now commonly use "if" as an alternative, as well as a condition: "The technical supervisor was uncertain if the new meter would read correctly under field conditions." However, using "whether" for alternatives is clearer for most readers:

The technical supervisor was uncertain whether the new meter would read correctly under field conditions.

Principle, principal. Do not confuse these two. As a noun, "principle" is a rule, natural law, fundamental truth, or guide to behaviour:

The geological principle of superposition says that younger rock strata are higher in the formation.

As an adjective, "principal" refers to the leading or most important person or thing:

The principal means of fire ground attack is the unit crew.

As a noun, "principal" can refer to the head of a school, or to a sum of money upon which interest is paid. Do not assume "principal" can refer only to a school principal.

Setup, set up. Many technical writers confuse the noun "setup" with the verb "set up"

Set up the online model with a setup of two-dimensional drawings.

You will find many other mistakes in English usage in both general and technical writing. Make your own lists of correct usage and watch for it in your documents.

Avoiding Errors in Spelling

Spelling errors can mar the effect of good writing and composition skills. Readers will tend to focus on spelling gaffes and miss the good qualities of your work. A good, college-level dictionary of Canadian English is essential, and the spell checker in your word-processing software is a great help in spotting errors. Microsoft Word XP comes with a powerful checker that can be set to Canadian English.

Remember that scientific and technical terms are not typically included in the software's list of English words and variations, so that you will have to add to the word processor's "dictionary" the terms that you will need when checking your own documents.

Remember also that the spell checker will not find every kind of error. It will find your spelling mistakes and typographical errors quickly and accurately, as long as you have not added incorrect words to its dictionary. However, you will make other mistakes, such as writing "controls" when you meant "control," or "form" when you meant "from." These "substitution" errors are now common in mass publications that use automated spell checkers. You must proofread drafts of your papers yourself to find these errors and correct them.

Good Spelling Habits

1. *Proofread all final drafts of documents for spelling errors.* Besides making your writing more correct, this practice identifies for you those words that you typically misspell. List words that you often miss, look them up in a good dictionary, and learn their spelling, pronunciation, and various meanings.

2. *Learn to pronounce words properly.* If you say a word without some of its vowels or consonants, then these omissions will often appear in your written version of the word. Two common examples are "environment" and "government"; if you say "enviroment" or "goverment," then you will likely misspell them.

3. *Learn the rules of spelling.* Knowing and following some basic conventions is an efficient way to correct spelling because English spellings are, for the most part, consistent. You may have to memorize some individual exceptions to a rule, but learning and following the rule first will avoid most mistakes and the waste of time that goes with correcting them or looking up in a dictionary every word you are unsure of. Start with the simple rules given below. Watch for others as you proofread and check your own spelling.

4. *Learn the meanings of words.* Homonyms, for example, sound the same but are spelled differently and have different meanings, such as "bare" and "bear." This also applies to words that are similar in sound and spelling but different in meaning, such as "affect" and "effect." Learning word meanings not only helps you spell more correctly but also increases the range, power, and precision of your writing.

Spelling Rules

The "i before e" rule. Many learned this rule in school as a mnemonic jingle: "*i* before *e*, except after *c*, or when sounded as *ay* as in neighbour and weigh." Apply the rule and mostly you will be correct: "receive," "conceit," "lenient," and so on. Two exceptions are "weird" and "seize."

Prefixes. When adding a prefix to a root word, write them together without a hyphen: "un + noticed = unnoticed." Note three common exceptions to this rule, each of which requires a hyphen between prefix and root: (1) root words that begin with capital letters, like "pre-Mousterian," (2) prefixes that end with a vowel added to root words that begin with a vowel, like "re-energize," and (3) "self" as a prefix, like "self-closing doors."

Suffix rule with final silent "e." When a root word ends in a silent "e" and the suffix begins with a consonant, retain the "e" in the root word, like "hopeful." When a root word ends in a silent "e" and the suffix begins with a vowel, drop the "e" from the root word, like "hoping."

Suffix rule for doubling final consonants. When a root word ends in a consonant, double the final consonant when the following three criteria apply: (1) the suffix begins with a vowel, (2) the final single consonant in the root word is preceded by a single vowel, and (3) the root word is accented on the last syllable, as in "occurred" and "beginning." In all other cases, the final consonant remains single, as in "interpreting."

Suffix rule for root words ending in "y." If the "y" is preceded by a consonant, as in "accompany," change the "y" to "i" before adding a suffix, as in "accompaniment," unless the suffix begins with "i," as in "worrying."

If the "y" is preceded by a vowel, as in "annoy," leave the "y" unchanged in front of a suffix, as in "annoying."

Plurals and possessives. Form regular plurals by adding "s" to the root word:

microchip–microchips; measurement–measurements

Plural forms of modern acronyms may add "s" or " 's":

CDs–CD's; SMCAs–SMCA's

Learn irregular plurals as you require them: "mouse–mice," "wolf–wolves," "focus–foci," "criterion–criteria," and so on. Form singular possessives by adding " 's": "the burner–the burner's," "the switch–the switch's."

This includes words of one syllable that end in "s" or an "s" sound:

Charles's theory; Marx's principles of economics

For words of two or more syllables ending in "s" or an "s" sound, add only an apostrophe: "conscience' sake," "Achilles' heel."

Do not use an apostrophe with possessive pronouns: "hers," "his," "its," "ours."

Form plural possessives by adding "s' ": the burners–the burners', the switches–the switches', the probes–the probes'.

Some common words are difficult to spell. Try to memorize the spellings in this list:

mischief	rhythm	privilege	receipt
benefit	occurrence	accommodation	manoeuvre
supersede	seize	occurred	parallel
prejudice	separate	benefit	lose

Canadian Spellings

"I am Canadian: hung between styles."—Dennis Lee (Kingdom of Absence XXI)

We spell words for historical, not logical, reasons. Time changes language and its spellings. The influence of American English in traditionally British Canada continues to grow here in the 21st century. Young people in southern Ontario no longer sit on a chesterfield; they sit on the couch. Where does that leave Canadians?

Working documents in Canada use a blend of British and American spellings. We recommend this blend for your report. We also recommend that you have and use a college-level dictionary of Canadian English. The following list sets out the main points of the Canadian style of spelling.

- Use British "-re," *not* American "-er" endings: "centre," theatre," *not* "center," "theater."
- Use British "cheque," *not* American "check" when referring to banking.
- Use British "-our," *not* American "-or" endings: "colour," "humour," *not* "color," "humor."
- Use American "z," *not* British "s": "organization," "itemize," *not* "organisation," "itemise."

Using Technical Conventions

Numbers

The scientific and technical writer must include many quantities in the text of a report. Learn the conventions of technical style to help you decide whether to write the quantity as a numeral or in words.

- Use numerals for quantities combined with units of measurement: 500 rpm.
- Use Arabic numerals for quantities of 10 or greater. If no other rule applies, write quantities of less than 10 in words.
- Use numerals for all quantities in a series; that is, a list of quantified items in a sentence: "They used 8 thinning saws, 241 litres of paint, and 215 units of stock."
- Use numerals for times of the day, days of the month, quantities of money, decimal expressions, and percentages.

- Do not begin a sentence with a numeral. Either write out the quantity in words or rewrite the sentence.
- Write approximations or indefinite measurements in words. The language must clearly indicate that an approximation is intended: "The final distribution figures included *approximately* forty more samples from the refinery." Use approximations sparingly.
- Write very large quantities in a form combining numerals and words: "$23.4 million."
- In compound-number adjectives, write out the first number or the shorter number, and use a numeral for the other: "twenty 10-cm trout."

Abbreviations

Abbreviate units of measurement, names for technical materials, and names of organizations when appropriate. Learn commonly accepted abbreviations for units of measurement in your field of study, and follow these rules:

- Abbreviate units of measurement following numerals denoting an exact quantity: 18 ha, 2700 L.
- Write abbreviations in the singular only: 2700 rpm.
- Use lower-case letters for abbreviations except for letters standing for proper nouns or adjectives, or by convention: 180 psi, 2400 Btu, 147.060 MHz.
- Do not use periods after abbreviations, except when the abbreviation spells a word.
- Do not use signs, or visual symbols, for abbreviations. Exceptions include conventional use, such as the percent symbol in 28% or the degrees symbol in 45° North latitude, and in tables and figures.
- Abbreviate names of organizations and nontechnical terms as long as the first mention gives the full title. Form these abbreviations without periods or spacing: American Radio Relay League (ARRL), fiberglass-reinforced plastic (FRP). Such abbreviations are called *acronyms*.

Hyphens

Hyphens are used to form compound words. The technical writer uses many compound expressions for technical concepts and measurements. Therefore, he or she must know when to hyphenate such compounds. However, a distinct difference between American and British usage of hyphens often creates confusion for the Canadian technical writer. The Canadian writer should generally follow British usage.

- Connect prefixes with the root word: pre + determined = predetermined, but note the following exceptions:
  a) to permit internal capitalization: pre-Cambrian.
  b) to aid pronunciation: re-allocate, re-form (meaning "to form again")
  c) "self": self-closing doors.
- Hyphenate compound adjectives but do not hyphenate compound nouns (a sodium-chloride solution, but sodium chloride; a flow-chart evaluation, but a flow chart).
- Hyphenate to avoid ambiguity; that is, suggesting two meanings simultaneously for the same expression: two-hundred-gallon drums, two hundred-gallon drums.
- Never hyphenate a compound formed with an adverb: a "badly managed operation." When uncertain, observe dictionary usage.

Applying Scientific Conventions

The International System of Units (SI)

The International System of Units is the accepted form of the metric system that is advocated by the Canadian Standards Associations (CSA) and is the official system of measurements for Canada.

Species Names

Give the scientific name, together with the authority for that name, following the first mention of any common name for a species. This eliminates confusion arising from the use of a variety of common names for the same species. After the first mention, the common name may be used alone. Note that the *abstract* is a separate report element and species names in the abstract therefore must include both common and scientific names.

Use capital letters in common names only for proper nouns; that is, those that name a person, place, or thing. Some examples follow.

- black spruce (*Picea mariana [Mill-] B.S.P.*)
- Norway spruce (*Picea abies [L.] Karst.*)
- bunchberry (*Comus canadensis L.*)
- spruce budworm (*Choristoneura fumiferana [Clem.]*)

Italicize the Latin words used in scientific names. Capitalize the generic name but not the specific. Consult taxonomies of species for the correct abbreviations for the names of authorities.

The formula for heat flow under these conditions is

$$Q = \frac{k(T_1 - T_2)At}{d} \tag{5}$$

where

Q	=	heat flow,
k	=	coefficient of thermal conductivity for the refractory material,
$T_1 - T_2$	=	temperature drop from hot face to cold face,
A	=	area of the wall,
t	=	time,
d	=	thickness of the wall.

Box 2.1: Format for Mathematical Equations

Mathematical Equations

- Space all mathematical equations beginning on a separate line from the text (see Box 2.1). Centre short equations, and begin longer ones flush with the left-hand margin and continue them, if necessary, on the second and successive lines indented two spaces.

- Number long equations and formulas in the text. Place a numeral in round brackets on the line below the equation and on the right margin of the page. When referring to the equation in the text, use the form "Equation (5)," where 5 is the equation number assigned in sequence through the report.

- Insert mathematical equations into your text using the Word XP Equation Editor. See Appendix B for a tutorial on the use of the Equation Editor.

CHECKLIST FOR THE REPORT-WRITING PROCESS

Review of the Steps in Report Writing

1. Gather all available research materials.
2. Plan before you write.
3. Draft the main points of the body.
4. Write your concluding sections.
5. Add your graphic presentations.
6. Write your introduction.
7. Compile the report elements.

Be sure that you have...

- Completed all necessary research
- Written research notes with clear paraphrases, including bibliographic information about sources; checked all secondary sources and cited them
- Organized information into a coherent, logical outline of headings
- Learned the computer's capabilities, especially word-processing software
- Drafted text using information from notes and followed outline
- Checked paraphrased secondary sources for originality; ensured all sources are cited and listed
- Listed accurate primary source information and measurements
- Written short, coherent paragraphs; written sections under logical, descriptive headings
- Revised and checked content, amount of detail, specific examples, transitions
- Ensured writing style is accessible, comprehensive, clear, correct
- Restricted use of passive voice within reasonable limits
- Used a formal tone (third person, past tense)
- Corrected grammatical errors (fragments, run-on sentences, agreements of subject–predicate and pronoun–antecedent, principal parts of verbs, misplaced or dangling modifiers)
- Corrected punctuation errors (comma, semicolon, colon)
- Corrected stylistic errors (pronoun references, diction, parallelism)
- Checked proper usage (accept/except, amount/number, data, effect/affect, if/whether, principle/principal)
- Checked spelling (proofread after spell checking, apply rules, record misspelled words, use Canadian spellings)
- Used correct technical conventions (use of numbers, abbreviations, and hyphens; correct metric (SI) units; scientific names of species; mathematical equations)

NE

Chapter 3

Formatting Your Report

Overview: This chapter describes how to format informal and formal reports, and how to set up illustrations to support your written descriptions of scientific and technical work. The discussion includes the physical form of a report, including page format and report elements, and the presentation sequence for information in a typical report.

REPORT STRUCTURE

According to semanticist Dr. S. I. Hayakawa, we can make only three kinds of statements with language: a report, an inference, and a judgment. A *report* is a statement of fact; it is verifiable. An *inference* is a conclusion we draw about what we do not know based on what we do know. A *judgment* is a statement of opinion about what we like or dislike. People often confuse these three, thinking that "This type of stadia rod is no good" is a statement of fact, when it is really an opinion. Structure your scientific or technical reports to separate and focus on each of these three kinds of statements in distinct, clearly labelled sections.

The body of a report contains descriptions of scientific and technical methods used to investigate problems or to gather data. It also describes the results of investigations, usually including tables of data. This is the reporting function essential to technical writing. Information must be accurate, complete, and presented in simple, clear language with appropriate tables and figures.

The body will also usually contain some analysis of the problem or data. These are inferences, conclusions that you draw based on evidence. Secondary sources and standard mathematical formulas, such as chi-square analysis, will help you reach accurate, logical conclusions. Reasoning through to a valid conclusion is never easy. Give yourself time to think and to rewrite the concluding report section several times. Your education and work experience will be invaluable in this process.

Recommendations may also be included at the end of the body. Do not include a section of recommendations unless your reader has specifically asked you to do so. Recommendations are your opinions; we all make judgments about our experiences. However, the reader of scientific and technical writing will consider only judgments based on carefully reported facts and logical conclusions. If you conduct your investigations carefully, write up the detail plainly, analyze the data correctly, and recommend an option that fits the data and analysis, your supervisor will be more likely to follow your recommendation.

INFORMAL REPORTS: LETTER AND MEMO REPORTS

The most common reports in the workplace are the shortest. Regular communication of scientific and technical information is essential to the effective operation of a business or government agency. Employees responsible for such information typically communicate it in the form of letters or memos, now sent routinely as e-mail, as well as short reports up to about ten pages. Write letters to people outside your company or agency; write memos to people inside your company or agency. Use e-mail for both sets of correspondence.

Letters

Letters have conventional parts. Learn them and learn how to set them up in your word-processing software. Box 3.1 shows the correct form of a letter in full block format, the most commonly used and easiest to write on your word processor.

The full block format begins each line on the left margin of your page. Other letter formats make use of indented paragraphs and right-justified blocks for the heading and complimentary closing with typed signature. These take longer to format, but have a traditional look preferred by some companies or agencies. Single-space the blocks of type in a letter. Leave blank lines between blocks; adjust this blank space after writing the letter to centre the text vertically on the page. Use a single, common typeface for the letter, such as Arial or Times New Roman.

- *Heading*. At the top of your page is the heading. The heading shows the address of the person sending the letter and the date when you wrote the letter. Do not include your name in the heading. Your name

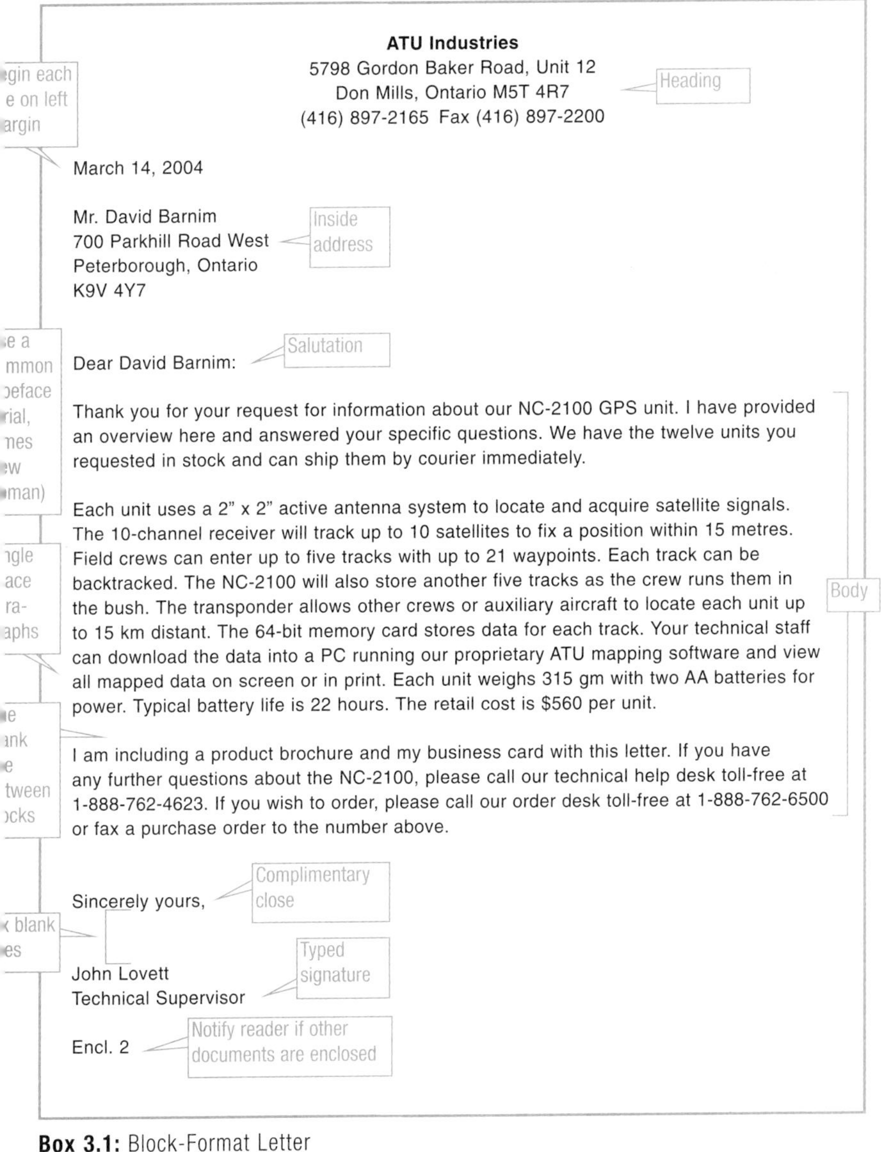

ATU Industries
5798 Gordon Baker Road, Unit 12
Don Mills, Ontario M5T 4R7
(416) 897-2165 Fax (416) 897-2200

March 14, 2004

Mr. David Barnim
700 Parkhill Road West
Peterborough, Ontario
K9V 4Y7

Dear David Barnim:

Thank you for your request for information about our NC-2100 GPS unit. I have provided an overview here and answered your specific questions. We have the twelve units you requested in stock and can ship them by courier immediately.

Each unit uses a 2" x 2" active antenna system to locate and acquire satellite signals. The 10-channel receiver will track up to 10 satellites to fix a position within 15 metres. Field crews can enter up to five tracks with up to 21 waypoints. Each track can be backtracked. The NC-2100 will also store another five tracks as the crew runs them in the bush. The transponder allows other crews or auxiliary aircraft to locate each unit up to 15 km distant. The 64-bit memory card stores data for each track. Your technical staff can download the data into a PC running our proprietary ATU mapping software and view all mapped data on screen or in print. Each unit weighs 315 gm with two AA batteries for power. Typical battery life is 22 hours. The retail cost is $560 per unit.

I am including a product brochure and my business card with this letter. If you have any further questions about the NC-2100, please call our technical help desk toll-free at 1-888-762-4623. If you wish to order, please call our order desk toll-free at 1-888-762-6500 or fax a purchase order to the number above.

Sincerely yours,

John Lovett
Technical Supervisor

Encl. 2

Box 3.1: Block-Format Letter

appears only at the bottom in the typed signature. In the example, the heading consists of the company letterhead with the contact information and the current date on a separate line below the letterhead. If you are not writing the letter from your desk at work, use your return address typed on the left margin, like this:

316 Ritson Road
Oshawa, ON
L1G 5P8

March 14, 2004

Put Canadian postal codes on a separate line below the street address. Write out the date in full; do not use numeric codes.

- *Inside address.* The block below the heading is the inside address. This address will appear on the envelope, and includes the name and street address of your correspondent. Check the details carefully, especially the name and title. Address your letter to a specific person whenever possible, someone who has the expertise and the authority to act on your letter. If no specific person is available, address it to a position: "Personnel Manager," "Network Manager," "Director of Information Services," and so on. Least desirable is the letter addressed to a large company; such letters, if answered at all, are often answered by someone not in a position to help you. If you must address a company, add an attention line below the inside address to direct the letter to a specific person.

- *Salutation.* The "Dear Somebody" part of the letter is the salutation. Salutations are conventional and must agree with the first line of the inside address. When addressing a specific person, use his or her name as you find it: "Dear Adrianna Taylor." When addressing a position, use "Dear Sir or Madam." When addressing a company, use "Ladies and Gentlemen." Punctuate the salutation of a business letter with a colon.

- *Body.* Type the body of the letter in single-spaced paragraphs beginning on the left margin. Leave a blank line between paragraphs. Letter paragraphs must be short.

- *Complimentary close.* After the letter body, type a complimentary close. Begin with a capital letter and end with a comma. Use conventional ones that most readers recognize: "Sincerely yours," "Yours truly," and so on. Leave about six blank lines for your signature, and type in your name as you wish others to use it. Include your position with your company or agency if this will help your reader to identify you and your concerns. If you intend to include other material with the letter in the mailing envelope, leave a blank line below your typed signature, and type the word "Enclosure" to signal the reader that something else is in the envelope. You may abbreviate this to "Encl." and add a numeral to indicate the number of enclosures.

Composing Letters

Letters always address a single issue. Do not attempt to compose a letter that deals with more than one major subject. Follow the correct writing process when composing a letter: consider your audience and purpose, gather all necessary information, sketch an outline of paragraphs, and compose the letter using short paragraphs with strong, clear topic sentences. Deciding on your purpose for writing the letter and expressing it clearly in the opening paragraph are the most important factors in writing a successful letter.

Write an opening paragraph that clearly states the purpose of the letter. Use action words: "to *review* the budget for the field instrumentation project," "to *summarize* the lab-testing data gathered to date," or "to *provide* the input you requested for staffing needs in July and August." Do not introduce yourself; your name and position will appear in the letter as the typed signature. Keep the opening paragraph short.

Write the details of the letter in the middle. Scientific and technical information about principles and processes can be complex. Therefore, group your details into focussed, logically connected topics, and write a short paragraph about each one.

Write a concluding paragraph that clearly indicates the response you wish from your reader. Business letters ask the reader to do something or think something. Know your audience and your reasons for writing the letter. Express your expectations clearly and courteously in the closing paragraph. Use action words: "Please send the requested information to the address above," "If you have concerns about the new data, please contact me by return mail," or "Please let me know your decision with regard to staffing as soon as the summer budget has been approved." Notice that these three sentences economically accomplish the objectives of a business letter: purpose stated, business details, action required. When you have finished composing, formatting, and revising the letter, print a copy and sign it in the space between the complimentary close and the typed signature.

Letters can be quite brief. A follow-up letter from ATU Industries might look like the example in Box 3.2.

Memos

Organizations use memos for internal communication. Memos provide permanent file records of activity and decisions made, and communicate routine news to individuals throughout an organization, summarizing key information and suggesting action. Use memos for brief scientific or technical reports that include sections written under headings and graphic presentations of data.

The conventional form of the memo begins with four lines of text, each introduced with conventional guidewords, like the following:

To:
From:
Subject:
Date:

Write the body of a memo in block form, starting each line on the left margin. Use single spacing with a blank line between paragraphs, as in Box 3.3.

You may choose to sign a memo to lend weight to an important directive, and may wish to introduce a memo with the word "Memorandum" to identify the type of communication, including the company logo or personalized information. You may also wish to insert a graphic line below the guidewords to separate the heading visually from the body of the memo.

Composing Memos

Like letters, write memos on only one subject. Begin the memo with a short, direct statement of your purpose in writing. Use action words such as "to request," "to explain," "to authorize," and so on. Add any necessary explanations: "This request for a change in scheduling is the result of a shortage of parts at the main supplier's warehouse."

ATU Industries
5798 Gordon Baker Road, Unit 12
Don Mills, Ontario M5T 4R7
(416) 897-2165 Fax (416) 897-2200

March 31, 2004

Mr. David Barnim
700 Parkhill Road West
Peterborough, Ontario
K9V 4Y7

Dear Mr. Barnim:

Thank you for your order of March 23.

It will take four to six business days to pack and ship your GPS units by courier.

Payment is due in 30 days.

Sincerely yours,

John Lovett
Technical Supervisor

Box 3.2: Follow-up Letter

MEMORANDUM

TO: Gary Farrow, Manager, Lab Services
FROM: Leah McVeigh, Titration Lab Technician
SUBJECT: Revised Procedure for Equilibrium Titration
DATE: September 18, 2004

In line with our upgrade of water testing facilities, we have purchased a new resonance biosensor for equilibrium titration of water samples. The following is the revised procedure that we are using with the new biosensor. Preliminary results suggest that we are getting measurements that are more accurate for a wider range of potential pathogens.

The new procedure is as follows:

1. Arrange a closed loop with two micro flex loops and two stoppered flasks.
2. Connect the external syringe pump to one of the stoppered flasks.
3. Set up the biosensor unit and insert the two sensor spots at opposite ends of the closed loop. One sensor binds interactive molecules in the fluid passing over it, while the other serves as a reference value for total units.
4. Inject a 15-cc water sample.
5. Activate the syringe pump and check for continuous circulation of the sample.
6. Activate the biosensor and record differential readings from the display.
7. Draw off a 2 cc sample and perform a standard stepwise titration.
8. Multiply the titration results by the biosensor readings.

The new sensor gives readings that are more consistent over a wider range of temperatures and water conditions and requires simpler operating steps. If you wish, we will arrange a demonstration in the lab at your convenience. A report comparing titration results with the old and new methods is being prepared. Please call if you have any questions.

Box 3.3: Memo

The middle part of a technical memo can be several paragraphs long, describing specific details of processes and products. For this reason, technical memos can have two or three pages. Insert a descriptive heading at the beginning of each major topic covered in multi-page memos. Check your paragraphs to make sure they are short, describe only one topic, and follow each other in a logical sequence.

Technical memos can also include illustrations when you need them: brief figures to explain processes or show how devices function, brief tables to summarize data or compare technical parameters. Use bulleted or numbered lists wherever possible. In the last paragraph of your memo, describe the action you want your reader to take, such as returning a questionnaire or discussing the memo's information with staff, and remember to provide contact information. Make sure your reader knows how you wish to be contacted and any deadlines you have associated with the requested action.

Organizations use specific types of memos to communicate in common workplace situations.

- A *directive memo* defines a policy or procedure for your readers to follow. Typically, you will send this memo as a supervisor to your staff. When composing the directive memo, explain first the reasons for the directive and then state it as a request: "Please submit all timesheets and invoices for this phase of the research before May 1." A polite tone works best.

- A *response to an inquiry* provides information requested by someone in your organization. Open with your purpose: to respond to a request for information. Summarize the content briefly. Then discuss the detail that is requested. Scientific and technical information can be difficult to explain, especially in a memo. Therefore, organize the content of your explanations and descriptions into short, logically sequenced paragraphs. Close with a request for a response, if needed, or an offer to assist with information at a future date.

- A *trip report* is a record of a business trip. Organizations require these to account for travel funds, to record observations made in the field, or to report on meetings held with clients and associates in other places. Report only important details.

- *Field and lab reports* provide supervisory staff with the results of inspection and maintenance procedures. The memo should include the purpose of the work, the problem or hypothesis that you investigated and the methods you used to do so, a description of the steps in the investigation, a summary of the results, and recommendations if required.

- A *request memo* asks someone in your organization for information. Start the memo with a direct request and explain the reasons for it. Most requests for scientific or technical information require detailed and specific answers. Therefore, it is best to present the areas of concern as a bulleted list of specific, detailed questions. Vague questions get vague, unhelpful answers. Close with a reasonable deadline when you need the information and the method by which you wish to receive the reply.

E-mail

E-mail is used for the electronic form of letters and memos, distributed over local intranets or the Internet from your workstation. E-mail messages are formal documents. Students become accustomed to informality in personal e-mail, including the use of short forms, capital letters, emoticons, and bad spelling and grammar; you must curb these habits in the workplace.

Communications dealing with scientific and technical subjects must be accurate. The rules of good writing apply. Any message you send reveals your personality and the quality of your work.

As many people have discovered to their horror, e-mail is not private. Messages are stored as files accessible by others. Network software can allow supervisors to see the messages you have sent and received. Inappropriate form or content, then, can hurt your career. Treat e-mail as you would a formal memo.

Composing E-mail Messages

- *The subject line:* E-mail software provides the standard guidewords used in memos. Most important of these is the subject line. Do not write vague or confusing subject lines. Most readers must prioritize hundreds of daily e-mails, separating spam and routine communication from more urgent workplace concerns. Your subject line is the only clue your reader has about the content and urgency of your message. To write a good subject line, write the e-mail first and then use the key words from your message to build a descriptive subject line.

- *Choosing your audience:* Send e-mail messages only to those people who need to receive them. Electronic address books and list serves make it easy to send a message to dozens of people, most of whom do not need the information. Be careful when replying to a list serve message. Clicking the reply button will send your e-mail to everyone on the list, even though you intend your reply only for the originator of the message.

- *Length:* Keep messages short. As a general rule, fill one screen (about 25 lines) for routine messages. Scientific and technical information will require more text.

- *Composing your e-mail:* For detailed e-mails on scientific and technical subjects, work out the paragraphs on a word processor and paste them into your e-mail software. For longer documents and graphics, attach text and graphic files to a short, explanatory message that identifies the purpose, content, and file format of each of the attachments.

- *Proofreading:* Proofread every e-mail you send. This is harder to do on a computer screen than on paper; technology pushes us to "get things done" faster without taking the necessary time to think through the issues, from content to logical sequence to correct language. This leads to miscommunication when the words do not express accurately what we intend to say. Miscommunication causes wasted effort and resources. The more important the message, the more time you need to write it. Always think twice about an e-mail before you click the "Send" icon.

Short Reports

The Latin roots of the word "report" mean "to carry back." The purpose of a short report is to carry back important information in a written form. The information you carry back to supervisors from lab or field investigations in the form of written reports is essential to their administrative work. Since you will likely know your supervisor and the purpose of your report, choosing the right content and the appropriate form for your report will be relatively simple.

Many lab and field reports begin as fill-in-the-blank forms. Gather information using appropriate lab or field procedures so that you can analyze results accurately. Recent developments in hand-held data recorders have adapted written forms to electronic files in order to manipulate them with common computer software.

Composing Short Reports

Short reports are usually between five and fifteen pages long in manuscript form. For reports shorter than five pages, write a two-page letter or memo. For reports longer than fifteen pages, consider adding more report elements to keep track of increasingly complex content. The section below on the format of a formal report lists and describes its elements. A short report requires few formal elements. Typical short reports that contain only primary research will include the following elements:

- Title page
- Abstract
- Introduction
- Body (divided into sections with descriptive headings)
- Conclusion
- Appendix

You will find complete descriptions of these report elements in the section on elements of the formal report beginning on page 73.

FORMAL REPORTS

Formal reports are longer than routine documents, usually 20 pages or more. They deal with complex scientific and technical issues; therefore, they require additional formal elements, like an index or glossary, which provide access to the information and promote understanding of the report's content. In addition, they require a letter or memorandum of transmittal to transmit responsibility for the contents of the report to a specific reader or readers.

Formatting Formal Reports

Format for reports is similar to the traditional form of manuscripts: double-spaced text on good-quality, 8½ by 11–inch paper. Before looking at details of page format, let us consider for a moment the use of word-processing technology to format report pages.

Word Processing

Format report pages using available computer technologies, including word-processing software and laser or ink-jet printing. Skillful keyboarding makes typing text more efficient. Learn the features of your word-processing software thoroughly; it will make your job of formatting the report much easier. Writing a report on a computer requires you to use both keyboarding and composition skills. Keyboarding can be quick and efficient—thinking through your writing issues is often not as quick. Do not let the computer's speed push you into preparing a document without the necessary revising, editing, and proofreading steps. Allow sufficient time to complete your research, to compose clear text at the keyboard, and to revise the text to make it readable. It also helps to ask a knowledgeable person to review the manuscript for you.

Spacing and Margins

Double-space your report. Use single-spacing in specific elements, such as the abstract, long quotations, vertical lists, and entries in the reference list. Business reports often use 1.5 spacing. Indent the first line of each paragraph about five spaces. Do not write block paragraphs without indentation.

Do not allow the first line of a paragraph to sit alone at the bottom of a page (orphan line) or the last line of a paragraph to sit alone at the top of a page (widow line), and do not allow main headings or subheadings to sit alone at the bottom of a page.

Use the default margins in your word-processing software. These are usually 2.5 to 3.8 cm wide (1 to 1.5 in.). Direct quotations of secondary sources longer than 75 words are usually set single-spaced within margins of 5 cm (2 in.) or wider. However, such long text passages are not desirable in scientific writing.

Normal spacing for punctuation is two spaces following periods, question marks, and colons, and one space following commas and semi-colons.

Typefaces

Use a standard serif or sans serif typeface such as Times New Roman or Arial. These are Microsoft True Type™ fonts that most word-processing software can read and print easily. They are also easy for your readers to see. Typefaces should be 10, 11, or 12 points—anything smaller is difficult to read; anything larger uses more pages than necessary. Note that actual typeface size varies with design and leading. Text in 12-pt Garamond typeface takes less room than the same text using 12-pt Century Schoolbook typeface.

Headings

Organize your content into an outline of descriptive headings with at least two levels. If you can group any section of your report into a third level of headings, do so. This helps both you and your reader to understand the content better. (Look at the headings in the sample student report in Appendix C.) Divide your content into three or four main topic areas, and use the key words from these topic areas to create descriptive headings. In this way, a report of moderate length will have about six main headings and a series of subheadings.

Two of the main headings your report should include are Introduction and Conclusion. In the workplace, reports can be longer than the typical 2000-word college paper. Such reports will require more organization and more main headings. In the body of the report, place the Introduction heading at the top of the first page. Place the other headings in order as you write, following the preceding paragraphs of text. Do not place each new main heading on a new page; this wastes paper and adds nothing to the layout of the document. Headings are your reader's first and most important access point, so set the headings apart from the text using simple graphic levels of hierarchy. Use the following system:

MAIN HEADING

Subheading

Third-level Heading

Boldfaced headings stand out clearly from the body typeface. To differentiate headings further from body text, use an alternate typeface close in design to the base font in the report. An example would be the use of the sans serif Tahoma font for headings with the sans serif Arial base font.

When you refer frequently to other report sections with your report text, you may number report headings with the decimal system of numbering, illustrated below.

1.0 MAIN HEADING

1.1 Subheading

1.1.1 Third-level Heading

Note that heading styles, including design and numbering, vary considerably. Consult written guidelines or style manuals designed for documents submitted in your specific situation.

Page Numbering

Number the pages of the front matter if you have included three elements or more. Do not place a number on the title page, although it is considered the first page in the numbering system. In the front matter, use lowercase Roman numerals centred at the bottoms of the pages. In the report body, use Arabic numerals in the upper right corner. Begin page numbering with the first page of text (Introduction). It is not necessary to alternate page numbers between the upper right and left corners. Print reports on one side of the page only. Appendix B describes how to number pages in Microsoft Word XP.

Report Presentation

College professors will usually accept reports and essays that are stapled in the upper left corner. Be sure you have assembled the elements of the report in the correct order before stapling, and do not fold your report. Some college courses and most workplace situations will require more formal types of report presentation (e.g., envelopes and report covers). Add a label to each envelope or report cover showing the report title, the author, the person or group to whom you are submitting the report, and the deadline date.

Report covers may be heavy paper, vinyl, or clear plastic. Some college instructors and workplace readers will accept clear plastic covers with a plastic wedge spine or report covers that require three-hole punching your report pages. The most acceptable format for report covers, though more expensive, is spiral binding, which allows pages to lie flat.

Elements of Formal Reports

For examples of report element format, consult Appendix C. Begin each element of the front and back matter on a new page, but place the elements of the body on continuous pages.

Title Page

Centre the text on the title page between the margins. Space the title page elements vertically, equidistant from each other. Leave at least four blank lines from the top margin to the title at the top and four blank lines from the bottom margin to the date at the bottom (see Box 3.4).

THE DEVELOPMENT OF MAYAN EPIGRAPHY
AND ITS IMPACT ON INTERPRETATION OF CLASSIC
MAYA CULTURE

By

Jonathan L. White

To

Professor C. L. Gulston

October 23, 2004

Box 3.4: Title Page

Write the following four required elements on the title page in this order:

1. *Title:* In capital letters, write a report title that is long enough to be fully descriptive of the content, including all key words.
2. *Author:* Under a "By" line, write the name or names of the person or persons responsible for the paper. This may include their formal titles and departments.
3. *Destination:* Under a "To" line, write the name of the person or persons responsible for reading the report and who take responsibility for its contents.
4. *Date:* Write the deadline date (or the current date, if you are submitting the report before the deadline date). Write the date out in full without using numeric coding.

If required, the title page may contain information such as a company logo or report serial number, but do not place graphic images on the title page.

Table of Contents

Centre the heading TABLE OF CONTENTS in bold capital letters (see Box 3.5). In a column on the left, add the elements of the front matter that come after the contents page, the descriptive headings in the body, and the reference list. Include appendices in the list if you have included any in your report. Single-space your entries. In a column on the right, opposite each heading, list the page number where the reader can find that heading.

Do not list both the starting and ending page numbers for each section.

Insert leader dots connecting the headings with their page numbers (Appendix B describes how to do this in Microsoft Word XP). Add the heading "Page" over the page number column.

List of Illustrations

Centre the heading LIST OF ILLUSTRATIONS in bold capital letters (see Box 3.6). You may also call this element "Illustrations" or "List of Tables and Figures," the latter for longer lists that are subdivided into tables and figures.

In a column on the left, add the figure or table number and title for each of the illustrations in the body of your report. Do not include illustrations that are placed in the appendix. Double-space your entries. In a column on

TABLE OF CONTENTS

Page

ii

Box 3.5: Table of Contents

LIST OF ILLUSTRATIONS

Page

iii

Box 3.6: List of Illustrations

the right, opposite each entry, list the page number where your reader can find that figure or table. Insert leader dots connecting the headings with their page numbers. Add the heading "Page" over the page-number column.

Abstract

A descriptive abstract is a brief summary of the report's content, about 250 words long (see Box 3.7). It is a separate element of the report. Place the abstract in the report immediately following the list of illustrations. Centre the title ABSTRACT at the top of the page. Single-space the abstract text below the heading.

The abstract summarizes the following content of the report:

- Research goal or goals; the questions answered by your investigation.
- Research methods, including both fieldwork and literature searches.
- Summary of results, including the main topic covered in your report and any analysis of lab or fieldwork. Do not refer to your data or discuss the details here.
- Main conclusions, including reflection on the material presented in the report.

The abstract must be an independent summary that does not require the reader to look up references in the text of the report. An informative abstract may be longer, up to 400 words, including a summary of the report's recommendations.

Report Body

Divide your report into separate sections, using boldfaced headings and subheadings to identify the content of each section. Write topic headings; do not use sentences or questions for headings.

Reports have a beginning, a middle, and an end. Give the beginning section the title INTRODUCTION to set the reader's expectations. The middle of the report is your account of your investigations, divided into a logical sequence of sections; group your sections into three or four main headings, using a formal outline. Divide the main headings into subheadings and third-level headings where appropriate. The concluding section presents analysis of the data from field or lab work.

ABSTRACT

The purpose of this investigation is to summarize current knowledge of Maya writing, the forms that Maya writing takes, and the history of the decipherment of Maya writing. To illustrate these points, this paper describes Maya glyphs found at three major sites of the Classic Maya. Sources for this paper include books by Michael Coe and Norman Hammond on Maya history, writing, and epigraphy, as well as a number of Internet web sites.

Four Maya codices currently exist, the largest being the Madrid Codex. The majority of glyphs are found on limestone stelae at the major city sites of the Classic Maya. In addition, glyphs are found in wall murals and specialized structures like the hieroglyphic stair at Copan. Glyphs include symbols for numerals, people, and language phonemes.

Tatiana Proskouriakoff discovered the key to deciphering Maya glyphs in the 1950's. Recent work by the late Linda Schele and David Stuart has increased our understanding of Maya writing to include about 85% of extant glyphs. The glyphs on monuments at Copan, Tikal, and Palenque all tell the history of the rulers of those cities, their families, and their personal achievements. The deciphering of Maya texts has significantly altered our interpretation of classic Maya civilization.

iv

Box 3.7: Abstract

Introduction

Start numbering pages with Arabic numerals in the upper right corner of your introduction. Do not put the report title on the first page of the body. The purpose of the introduction is to answer your reader's questions about the report's purpose, how information was gathered, and how much information is included. You can provide answers to each of these questions in a separate subsection of your introduction. The following descriptive subheadings identify typical reader concerns about the report.

- *Purpose and scope.* Describe the primary goals of your research, adding a description of any secondary goals, such as providing information to a supervisor, or reporting to a safety committee, or completing requirements for a college course. Describe the scope and limitations of your investigation (remembering that time and resources limit lab or fieldwork), including the specific questions your work addresses. Time and resources also limit library research. Describe the specific topics for which you have research information. Include a plan of development describing the main topics covered by the report.

- *Methodology.* Describe the field or lab techniques used to gather your data. Most of these are standard techniques recognizable by most scientific and technical readers in your field. Therefore, you need only name the technique. If you have modified the technique to suit the goals, study area, or instrumentation available, indicate these modifications here.

- *Literature review.* Describe and evaluate the secondary sources used, and describe how they contributed to the final report. Organize your description in more- to less-important order, starting with the sources that were most authoritative, up-to-date, and complete. It is not necessary to describe all the sources that you read, such as general background information that is not included in the report, but use correct documentation (see Chapter 5).

- *Background.* Here you can discuss state-of-the-art techniques or equipment used in the work, the personnel involved, the weather conditions, and so on. If you are reporting extensive fieldwork, include a section called Study Area, with a map and a description orienting the reader to the study area.

Keep the sections of your introduction brief. You can discuss important points in detail in later sections. The purpose of the introduction is to give your reader an overview of the report, a mental roadmap to follow as he or she reads. Boxes 3.8 and 3.9 provide illustrations of parts of the introduction.

Report Content

The sections after the introduction describe the new information you have developed through primary and/or secondary research. Write a formal outline of headings before attempting to write a first draft of these sections. It will speed your work and organize it logically for your reader. It is much easier to write organized, complete report drafts from an outline and a set of research notes than it is to develop a text directly on your word processor. The longer your report, the more levels of organization you will need in your headings. For a complete discussion of organizing information, see Chapter 1 on report research and organization.

Each section should completely describe a specific aspect or phase of your investigation. Use short topic paragraphs to introduce each section, reminding the reader of its purpose and relationship to the overall goal, then summarize the section and indicate its importance in the final paragraph. For a more complete discussion of report writing style, see Chapter 2.

Conclusion

Scientific reports and technical papers differ in their approach to conclusions. Scientific reports describe field and lab methodology first, and then report the data under a heading called Results. They present data in one or more tables, and the text describes the important features. The analysis of the data appears under a heading called Discussion. This section analyzes the data sets and often includes charts, graphs, or other visual presentations. In science, the heading CONCLUSION suggests a new scientific principle or discovery arising from the experimental data. However, for technical readers the heading CONCLUSION usually indicates only a summary and preliminary analysis of field or lab data.

College students write most term papers entirely from secondary research. Their purpose is to summarize information and ideas from the published works of recognized authorities in a clear, readable form, and to document those sources thoroughly. Therefore, the conclusion in a term paper is shorter and less emphatic than in a technical report that analyzes lab work or fieldwork. The conclusion of a term paper reflects on the content by summarizing the

1

INTRODUCTION

Purpose and Scope

The purpose of this investigation is to summarize current knowledge of Maya hieroglyphic writing, the forms that Maya writing takes, and the history of the decipherment of Maya writing. To illustrate these points, this paper describes Maya glyphs found at three major sites of the Classic Maya.

The scope of the investigation is limited to the significant details of Maya written texts and the history of efforts to decipher them. This includes significant changes in our view of Maya history and culture that have come about because of deciphering Maya writing.

Box 3.8: Introduction

2

Review of Literature

The most helpful sources were Coe (1992), Hammond (1982), and Schele (1998). Written with Peter Mathews, Schele's book explains the key elements of Maya writing and how they were discovered. Coe, an archaeologist and epigrapher, discusses the contributions made to Mayan epigraphy by scholars of the 20th century. Hammond, a British archaeologist who worked in Belize, gives a good overview of Maya history and culture.

The GBonline web site provided information about Maya codices and a timeline of the rise and fall of Maya civilization. David Stuart's online article (1996) is a comprehensive decoding of the history of the Maya rulers of Copan found in the glyphs on Altar Q.

Box 3.9: Review of Literature

most important ideas and applying them to the central question or purpose stated in the introduction (see Box 3.10). It may also reflect the writer's own thoughts or ideas, stated clearly and concisely.

List of References (Literature Cited)

The *List of References* is a separate element, and is placed on a separate page. It is the last numbered page in your report. Place the title in bold capital letters centred at the top. Choose the heading that is appropriate for your list of published sources.

A *Literature Cited or Works Cited* list includes only published works that you have paraphrased or quoted directly in the report body, using the correct form of citation.

A *List of References* includes not only published works that you have paraphrased or quoted directly in the report body, but also uncited sources. This type of material usually includes general reference works such as dictionaries, encyclopaedias, taxonomies of species, and textbooks. A student writer will typically use these resources in the early phases of research to focus the investigation and select appropriate topic areas for further study. In some cases, the writer may need to understand scientific or technical principles, processes, and practices that underlie the subject of investigation. These are included in a List of References for readers who are likely to want more information than just the cited sources. (See Chapter 5 for examples of documentation styles.)

Glossary

Write a glossary if you have used a large number of scientific and technical terms in the text of your report. Write a definition for each term in simple language, providing appropriate examples, and list the terms in alphabetical order. The glossary may also be placed in the front matter after the illustrations page. Shorter reports typically contain only five or six terms that your reader may not know. In this case, omit the glossary from your report but be sure to write clear definitions for these terms in the text when you first introduce them.

Appendix

The appendix is a separate element at the end of the report. It contains materials that support the descriptions in the text but are not essential to the text. This judgment about where to place supporting materials belongs to the writer. In general, the writer should include supporting materials in the body

13

CONCLUSION

The development of Maya writing closely parallels the development of Maya civilization. A good example is the shift in the early Classic period from the written forms used for the codices to the forms used on permanent monuments.

Sir Eric Thompson made the first translations of Maya glyphs, but these were few and led to misconceptions about Maya civilization as a passive culture of priests and astronomers. Tatiana Proskouriakoff found an important key in the use of name and city glyphs within Mayan "cartouches." She was thus able to translate most of the monument glyphs at Tikal and Palenque. These proved to be written histories of their warrior kings. Linda Schele extended Proskouriakoff's work to include the codices and many stelae.

Box 3.10: Conclusion

of the report when such materials are useful or essential to the reader's understanding of the written description. All other materials belong in the appendix. Appendices may include field or journal notes, tables of raw field or lab data, graphic material like sketch maps or seismograms, photograph or satellite image series, or copies of correspondence or other supporting documents.

Begin a new page immediately following the reference list with the single word APPENDIX centred in bold capital letters. Do not number the appendix pages. In appendices longer than ten pages, add a table of contents immediately following the APPENDIX title page. Group related types of supporting materials in separate appendices, such as photos or computer source code, and place a bold, capitalized heading centred at the top of the first page of each new appendix. Identify your appendices or groups of appendices with capital letters: APPENDIX A, APPENDIX B, and so on.

REPORT GRAPHIC FORMAT

You will use two types of report graphics: tables and figures. *Tables* display raw data from your lab and fieldwork; *figures* summarize data and show data trends—each type of figure has its own message. Your reader depends on graphics for the key information in your report. Learn the strengths and weaknesses of each form of graphic communication. Graphic literacy is as important to the scientific and technical writer as language literacy.

Graphic Placement

Choose your graphics carefully. Is the information important to your reader? Does it accomplish your purpose? If—and only if—the answer is yes, include the graphic in your report. A scientific or technical report is no place for graphic ornamentation.

If you have written a description of methods or instrumentation, consider adding a table or figure to summarize the data or to improve your reader's understanding of processes. Place these graphics on the page accompanying your written descriptions. If you have tables or figures that are associated with the subject of your report but are not discussed directly in your text, place them in an appendix. Examples of this kind of information might include tables of daily instrument readings, satellite photographs of study areas, or schematics of instrument circuits.

Graphic Integration with Text

Make good decisions about where to introduce your reader to a graphic in the flow of your description and analysis of work. Support the graphic by explaining its importance in your report's findings. Without this kind of integration, readers will lose your meaning; your text will be harder to understand without the graphic, and the graphic will lose its importance without the context of your discussion. Do not stuff all of your graphics into an appendix; deliver them to your reader at the right spot in your report.

Place a graphic immediately following the first text passage that discusses the information or idea it illustrates. Refer the reader to the graphic with a textual reference in parentheses, like this: (see Figure 1). Figure 1 will follow immediately on the same page, or, if it is a large graphic, on the following page. Use the correct commands in your word-processing software to wrap text around the graphic. You may discuss aspects of this graphic in later sections of your report; it is easier for your reader to refer back to a graphic that has already been introduced than to imagine one they have not yet seen.

Tables

Table Number and Title

Accessibility of information is essential in a report. The following rules allow your reader easy access to the tables in your report. Give each table a table number and descriptive title (see Box 3.11). Place these *above* the table. Use Arabic numerals for table numbers. Number your tables in sequence from the beginning of your text. Include each table's number, title, and page number on the list of illustrations contained in your report's front matter.

Begin the title block for the table on the left margin, single-spaced. You may add short sentences of explanation after the title. If the units of measurement are the same for all data, add them in parentheses after the title block. If you borrowed the data in the table from a secondary source, place an appropriate citation at the end of the title block. For more on citations, consult Chapter 5 on report documentation.

Table Format

Separate the title block from the vertical column headings with a heavy line or a lighter double line. Separate the vertical column headings from the data with a light line. If text or a footnote follows the table on the same page, separate the table from these lines of text with a light line. Do not enclose the table with a border.

You can create a table by importing a spreadsheet into your word processor, or by setting table columns directly on a word-processed page using the appropriate tab stops. In all cases, the finished table must have the correct scientific format. The normal format for tables is the *portrait* orientation; that is, with the shorter side of the page at the top. You can print tables that are too wide for portrait printing in the *landscape* format, with the longer side of the page at the top. Landscape tables must be rotated 90 degrees to the left, leaving the bottom of the table at the right-hand margin of the page.

Table 1. Frequency Distribution of Ceramics at El Pilar by Maya Period (Wernecke, 1993)

Period	Frequency	Percentage
Preclassic	4	1.5
Middle Preclassic	1	0.4
Late Preclassic	10	3.8
Early Classic	26	9.8
Late Classic	65	24.2
Terminal Classic	1	0.4
Indeterminate	158	59.8
Total	264	100.0

Box 3.11: Table Format

Microsoft Word XP has a table function that allows you to enter text and data quickly in a customizable format. However, the grid format was designed for business reporting and is not suitable for scientific and technical reporting. The same is true of the table templates available using the Autoformat command. To create an appropriate table, set tab stops for your columns and use the graphic line function to insert horizontal lines.

For another example of a table, see the student report in Appendix C.

Figures

Figure Number and Title

Give each figure a number and a descriptive title (see Box 3.12). Place these *below* the figure. Use Arabic numerals for figure numbers. Number figures in sequence from the beginning of your text. List each figure number and title in your list of illustrations, opposite the appropriate page number.

Begin the figure number and title on the left margin of the page, single-spaced. You may add a *caption*, or short sentences of explanation, after the title. When you have borrowed the figure or the data in the figure from a secondary source, place an appropriate citation of that source at the end of the title block (for correct citation forms, see Chapter 5). For small figures with text wrapped around them, begin the title block on the left margin of the figure itself. For text wrapping and figure format in Microsoft Word, see the tutorials in Appendix B.

Figure Format

Place large figures on separate, numbered pages immediately following the first reference to them in the report text. Place smaller figures on the same page as the text, using the text-wrapping features of your word-processing software. Appendix B shows how to wrap text around figures in Microsoft Word XP.

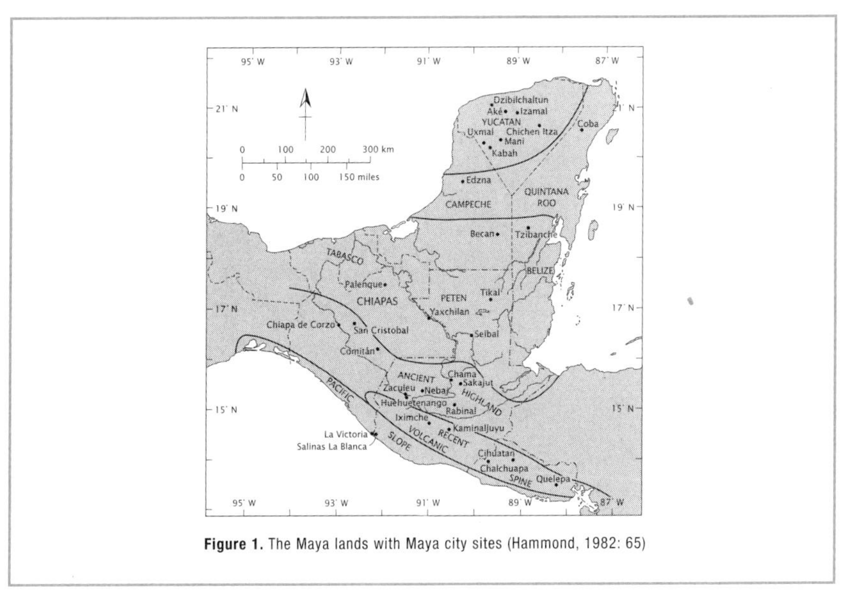

Figure 1. The Maya lands with Maya city sites (Hammond, 1982: 65)

Box 3.12: Figure Format

CHECKLIST FOR FORMATTING YOUR REPORT

Be sure that you have answered the following questions:

- Have you separated reports, inferences, and judgments into different sections of your report?
- Is your letter correctly formatted (heading, inside address, salutation, body, complimentary close)?
- Have you composed a letter that deals directly with only one issue?
- Are your letter's paragraphs short and sequenced as purpose stated, business details, and action required?
- Is your memo correctly formatted (guidewords, single-spaced paragraphs, a single topic)?
- What kind of memo is required? Does it fit a common writing situation? What is its structure?
- Have you written a descriptive subject line for your e-mail message?
- Have you sent your e-mail message only to those who need the information?
- Did you proofread your e-mail message carefully?
- Which formal elements are required for your short report?
- Did you check your report's spacing and margins, typefaces, headings, and page numbering?
- Is your report stapled, in an envelope, with report covers, or spiral bound?
- Have you checked each formal element (title page, table of contents, list of illustrations, abstract, body, reference list, glossary, appendix) for content and form?
- Did you choose graphics that communicate essential information to the reader? Are they placed in the best possible location in your report? Are they properly integrated with your text? Do all tables have numbers, titles, proper format, column headings, and units of measurement? Do all figures have numbers, titles, and proper format?

Chapter 4

Types of Reports

Overview: This chapter describes seven important types of workplace documents: instructions and manuals, progress reports, field trip reports, research reports, proposals, feasibility reports, and government reports.

INSTRUCTIONS AND MANUALS

A set of instructions is a common type of technical document, along with its longer form, the manual. You will write instructions for the operator—the person who will be doing the procedure you describe—with the purpose of allowing him or her to perform the procedure successfully.

Instructions vary in length from a single sheet to a dozen pages or more. You can also include instructions as a section within a longer scientific or technical report. In either case, good instructions will explain the proper steps in operating equipment accurately and safely or in performing lab or field tests.

Keys to Writing Successful Instructions

You will find that writing instructions for others about scientific and technical procedures that you are familiar with can be both satisfying and frustrating. Passing along your expertise to others satisfies professional and personal needs. However, technicians often find writing difficult. Words and illustrations in a set of instructions must appear in just the right combinations to produce a reasonable chance of the operator being able to perform the procedure successfully. Before you write, keep in mind the following keys to writing a successful set of instructions:

- *Keep it simple.* Wordy descriptions and elaborate examples prevent the reader from seeing clearly what to do next. Get directly to the appropriate actions in the next step.

- *Know the procedure.* If you are not completely familiar with all the steps, consult people who have the information you need.

- *Put yourself in the reader's place.* Most readers of technical instructions have some familiarity with basic principles and practices, but some do not. It is always better to give complete information to the operator than to assume he or she knows what to do. Remember how it felt to ride a bike for the first time? Your reader will feel awkward and uncomfortable the first two or three times through the procedure. Provide content that will direct him or her to a successful completion. This includes tests to determine whether the procedure was indeed successful.

- *Visualize the procedure in detail.* Then choose illustrations for your instructions that will clearly show your reader how to perform each step and what the outcome of that step should look like.

- *Test your instructions on readers.* No matter how well we express, organize, and support our descriptions with examples, we find omissions and areas that are not clear in a set of instructions. Ask a variety of people to read your instructions and, if possible, to perform the procedure, especially those who might actually be required to do so. This will provide you with valuable feedback about sections in your instructions that require more explanation or an additional diagram.

- *Consider your audience.* Your reader's level of expertise is critical to the success of your instructions. It is easy for a scientist or technician to forget his or her own level of expertise and assume the operator knows how to do simple subroutines. Never assume that the operator who reads your instructions will know how to do something. When in doubt, write it out.

Organizing Your Instructions

Studies of technicians who have to read and follow written instructions show that they typically do so after they have started the procedure and run into trouble. It is often said, "If all else fails, read the manual." This means that the operator who reads your instructions will often be in trouble in the middle of the procedure, scanning for the right information to help him or her out of a jam. Therefore, you must organize your instruction steps in a logical sequence under descriptive headings so that your reader can easily find the steps that are a problem.

To organize your instructions, group the steps in the procedure into three or four general tasks, each with a common outcome. Explain the outcome and list the steps involved in that task. This overview gives your reader a clear idea of the expected result of each task and lets him or her know when the steps have achieved their goal.

Tools, instruments, and materials are essential to proper performance of a procedure. Be sure that each section of your instructions begins with a description of these so that the operator knows what to gather together before beginning the task. If instruments are complex, such as spectrophotometers or magnetometers, describe and illustrate the controls, their function, and their proper calibration. If this description is several pages long, give it a separate major heading.

Common tasks found in sets of instructions include the following:

- Unpacking and setup
- Installing and customizing
- Basic operating
- Routine maintenance
- Troubleshooting

Safety messages are essential in any instructions describing procedures that might cause harm to the operator or the equipment being used. Standard safety messages are classified into three levels of urgency:

1. Danger: an immediate threat to the operator, such as the use of a chainsaw.

2. Warning: a potential threat to the operator if the equipment is used improperly.

3. Caution: potential damage to the equipment if used improperly.

Safety messages must be easy to see. Set them apart from the text by leaving a blank line before and after the message, and use a bold heading to identify which type of safety message you are presenting, such as the following:

WARNING
Move the power switch on the thinning saw to the "Off" position and apply the brake before attempting to sharpen or change the saw blade.

Safety messages must be easy to read; text written in jargon or technical terminology will not be understood by every operator and can lead to serious injury. Recent lawsuits have led insurance companies to require their insured companies to implement maximum readability levels for safety messages and instructions in their products. Therefore, define your terms clearly. Revise your message into the clearest, most direct language possible.

Place the safety message *before* your description of the steps that are potentially hazardous. This will increase the likelihood that the operator will have read and understood the message before attempting such steps and will avoid injury or damage to the equipment.

Write your instructions in chronological order as they are normally performed, and keep detailed notes as you do the procedure. This keeps the steps in the correct order and allows you to anticipate operator errors. It will also identify steps that

- need more explanation or detail
- must be illustrated with photographs or diagrams
- require the operator to choose among alternative routines within the procedure (these must describe criteria for making the correct choice)

Content of Instructions

Introduction

The introduction to your instructions will include the overall purpose of the procedure. Describe the main tasks and how they contribute to the final outcome. If the procedure is based on a scientific principle or theory of instrument operation, describe it briefly. An operator who understands the goal of the procedure and its basic theory is more likely to perform the procedure successfully.

The introduction will also include any training that the operator requires to perform the procedure properly. This includes any necessary safety instruction and preparation of the work area. List all of the tools, instruments, and materials that the operator requires in a bulleted list, and include any other background information that an operator might need, such as a study area map, a theory of operation, or available emergency contact information.

Procedures

Group all of the steps in the procedure into a simple sequence of tasks. Give each task a descriptive heading. Number the steps in sequence under each task heading. Keep your description of each step simple, focussing upon single, completed actions. Where the operator must make a choice between alternate procedures, list the options first and describe the criteria by which the operator must make the choice.

State your instructions in the imperative mood; for example: "Turn the edge clamp into place with the Allen key." Do not use the passive voice: "The edge clamp is turned into place with the Allen key." Except in short, one-page informal instructions, do not use "recipe style," which drops English articles (the/a/an): "Turn edge clamp into place with Allen key."

Integrate your illustrations carefully with the text. Good graphics are essential to the success of a set of instructions. Place appropriate graphics on pages with steps that require illustration. Wrap your text around small graphics, and set larger ones immediately above the steps that they illustrate; show the picture and then describe.

Conclusion

Conclude with necessary reminders to the operator:

- maintenance tips and schedules
- troubleshooting checklists
- steps to test the success of the procedure.

If your instructions are part of a longer report, write a short section that summarizes how the procedure effectively achieves its goal and links the procedure to the next major topic in the report.

Manuals

Manuals use generally the same format and structure as instructions. They are longer and include descriptions of instruments and complex processes that include more detail. They represent a greater investment of the writer's time and the company's resources. Manuals are usually collaborative efforts of people with specific expertise in writing and in scientific and technical procedures. The need for clearly written manuals is urgent because of increasingly competitive markets and the complexity of modern instrumentation. A good scientist or technician who can also write well is a valuable asset to a company that produces technical manuals.

Manuals have multiple audiences: the same manual may be read and used by operators, consulting engineers and technicians, corporate buyers, and repair technicians. They also have multiple purposes: step-by-step instruction, explaining operating principles, and reassuring the user. This requires multiple approaches to writing in different sections of the manual. It also means writing in the clearest possible English to appeal to a wide range of readers. Technical accuracy is essential in a manual; you will have to revise your text extensively and field-test the procedures you describe.

In the manual's front matter, write an overview of the contents and explain to the reader how to use the manual effectively. Identify typical users; the components of the products the manual describes, such as test instruments; and the purpose of the procedures it describes. Organize your content under logical, descriptive headings and lay out this plan for the reader in the introduction. Each reader of a manual is interested in only the specific sections that meet his or her needs. Therefore, organize and clearly label all sections of your manual.

Divide the content of a manual into chapters that are organized into labelled sections. Write in a clear, formal style and provide detailed illustrations. Tailor your content for specific audiences (such as diagnostic tests to check the operation of instrument assemblies for a field technician), and choose patterns of organization that suit your content.

Conclude a manual with necessary reference material for the user: a glossary listing alphabetically all scientific and technical terms, an index listing alphabetically all key words and headings by page number, or an appendix presenting data sets in appropriate ways such as graphs, reference charts, maps, or test checklists.

PROGRESS REPORTS

You will write progress reports as a project supervisor, usually for a middle manager. The manager, in turn, may summarize a number of progress reports in a monthly brief to senior management, and may pass along progress reports to a client who is paying for the contract. The purpose of the progress report is to update management and/or clients on project tasks completed and those remaining.

Managers need the information in your progress report to record your project's operational details, to evaluate it, and to account for the time and money you spent on the work. Regular progress reports allow managers to make mid-project decisions: to adjust timetables, to re-allocate budgets, or to reschedule supplies and equipment. They allow you to provide your manager with current information on the project and options for dealing with unanticipated technical problems and costs. Full, detailed progress reports meet senior management's needs for planning immediate and future company projects and for completing current projects on time and on budget.

Progress reports also satisfy the project supervisor's needs to fulfill his or her duties on the current project; these duties include keeping company management informed as the project moves forward. Fulfilling project duties in a timely and efficient manner is career development for the project supervisor. Good progress reports also allow the supervisor to more quickly write better completion reports at the end of each project. The form of the completion report is similar to the progress report; its detail is mostly a summary of the project that is already set out in progress reports, and its evaluation of the project comes from the detailed knowledge of the technical problems and solutions found in the project's progress reports.

Progress-Report Schedule

The schedule for reporting comes from the circumstances of the project. The manager responsible for the project will assign reporting duties to the project supervisor. If the project has been undertaken as contract work for a client company or government agency, the contract will spell out a project timeline that includes a progress-report schedule.

Reports are usually required at specified time intervals: weekly, biweekly, monthly, and so on. Progress reports can also be required at the completion of major tasks or stages of the project that are identified in the project timeline.

For smaller projects, progress reports can be less formal; for example, a client or project manager may request a progress report from time to time. The project supervisor may decide to submit a progress report when unusual or unforeseen circumstances arise on the job.

Progress-Report Format

For smaller projects, write progress reports as letters (external) or memos (internal). Larger projects involving longer periods, more detailed work, and more company resources will require more formal report formats, either short or long depending on the amount of scientific and technical information in the report and the number of formal elements required to provide the reader with easy access to the data.

Write all progress reports for the project using the same format and pattern of organizing information. This consistency ensures uniform, complete data so that you can properly evaluate the project's ultimate successes and failures. It is also easier for your audience to read successive reports if they all appear in the same format.

Tips for Progress Reports

Gather the information you need for your report. Remember that a progress report looks back at what you have accomplished and looks forward toward what is to come. Report your activities on the project over the last reporting period, checking for completeness and accuracy. Then forecast your activities on the project for the next reporting period. Focus on work that you and your staff have planned and solutions that you intend to apply to current problems.

The basics of a progress report consist of time, tasks, and topics. Organize your descriptions of work completed and planned in chronological order. Divide the steps completed and planned into *tasks*, the allocation of the work effort. Any problems encountered or evaluations of progress are special topics also set aside in separate sections.

A project supervisor will keep good records. A daily log or journal of events and a file of paperwork containing letters, memos, invoices, and so on will provide the complete, accurate, and detailed information you require not only in a progress report, but also in the completion report at the end of the project.

Parts of the Progress Report

The following is a general description of typical progress-report elements and their sequence. Each reporting situation has its own needs and variations on the basic requirements for information. Adapt the format shown here to individual college or work situations by adding to, omitting from, or re-ordering the topic list.

1. Introduction
2. Work Completed
3. Work Remaining
4. Problems and Adjustments
5. Conclusions

An executive summary is often included, since you will submit the progress report most often to management.

1. *Introduction.* Describe the scope and purpose of your project. If your project is the result of a successful proposal, check that document for the goals and scope of the proposed work. These statements of goals and limits for project work are important because they will direct your onsite decisions about allocating resources. Each progress report should clearly re-state the goals of the project, and sometimes you will have to alter the goals mid-project because of delays or technical problems. Describe any such changes to the scope and purpose of the project so that your reader can compare the original plan with the revised one. Include the dates for the reporting period covered by the report and the major stages of the project completed up to the date of the progress report.

2. *Work completed.* In this section, describe all the tasks completed on the job during the reporting period. Explain how you accomplished the work in steps. Group long step sequences under descriptive topic headings. For each main task, record the completion date. Include in this section any costs not in the project budget. Describe and explain any equipment or personnel changes that you made. Above all, describe and explain any problems or delays that occurred on the project. These are important to the manager who reads your report. However, do not include minor irritations that have no effect on the project timeline or budget. Summarize major problems that do affect the project in this section. Leave a full description of your difficulties, with some options for a solution, to a separate section dealing with problems and adjustments. If you had no problems and all is well on your project, say so.

3. *Work remaining.* Describe the tasks that you have planned for the next reporting period; give details of the tasks that you feel you can reasonably complete. Be realistic in your estimates. On-the-job experience will help with these. When reporting your work planning, you will find yourself wanting to estimate more work done in less time in order to impress your reader, often your boss. A manager is more impressed by conservative estimates with a high probability

of completion than by inflated expectations that result in disappointment. Conclude this section by summarizing all tasks remaining on the project after the next reporting period and estimating the expected completion dates.

4. *Problems and adjustments.* Do not include this section in your progress report if you have not experienced problems or delays and do not plan any changes to the planned work schedule. However, most scientists and technicians know Murphy's Law: if something can go wrong, it will. Most projects will develop some sort of serious problem that requires either solving or avoiding by changing the project goals. If Murphy strikes your project, describe the problem in this section of your progress report and detail any efforts you made to solve it. Explain any changes that you are either recommending or implementing to the original schedule, scope and purpose, specific tasks, or cost estimates. Include only major obstacles in your report; do not describe minor irritations. (By the way, Edward A. Murphy Jr. is a real person and an engineer, and he does not like Murphy's Law as people know it. The real Murphy's Law states that if there are two ways of doing something, one of which is wrong, then someone will do it wrong. He meant it as a caution to engineers, technicians, and scientists who develop lab and field procedures to remember to account for human error when designing projects.)

5. *Conclusion.* Summarize the status of your project in terms of time and budget. Evaluate your progress to date in terms of the relative ease with which you have reached some of the project's objectives. Forecast future progress, and recommend changes to planned tasks, costs, and scheduling as needed.

Goals of the Progress Report

Avoid generalities—if you have made little progress, say so. Be clear, short, direct, and factual, and do not pad the report. Select information carefully for your progress report. Remind your reader of the project's primary purpose and secondary goals, if any. Then include only details of events that promoted or delayed the successful completion of the project's goals. In scientific and technical projects, do not present partial data. Preliminary results can be misleading.

What does progress mean? If this report is about your progress on the job, what is progress? It can certainly mean completion of a specific phase of the planned work. It can also mean completion of preparation steps for the planned work: development of procedures for sampling or testing, ordering equipment, negotiating with suppliers, or briefing project personnel on required equipment or procedures.

However, progress can also mean failure. We learn more through our mistakes than we do our successes. Scientists, engineers, and technologists know Murphy's Law well. When something goes wrong on the project, as it inevitably does, we learn. We adapt. We persist. We overcome. Interim failure leads to results.

Repetition of information and format from one progress report to another is inevitable and desirable. Readers need reminders of the project's purpose and scope, and of the project's past progress, the work completed to date. Use your word processor. It is an invaluable aid in laying out the format and headings of your next progress report, the sequence of information required by your reader, the plan and purpose of your project, and the details of past progress that you would summarize in the next progress report.

Evaluate the progress on your project in the final section of your report. Answer management's key questions:

- Are we on schedule?
- Are we on budget?
- Are there any major problems arising?

Use illustrations such as graphs, charts, and photographs to show your progress toward the project's goals over time.

FIELD TRIP REPORTS

A field trip report records the process and the results of extensive field investigations, often in remote areas of Canada, in industries like forestry, geology, and environmental assessment. The audience for the field trip report is usually the investigator's immediate supervisor. Because this type of report contains detailed information and scientific data, it usually appears in formal technical report format with all the major elements of front and back matter.

Parts of the Field Trip Report

Divide the introduction to your field trip report into the following sections:

- *Aims.* Describe the primary goals of the fieldwork and any secondary goals. Use action words to describe these goals: "to observe conditions," "to collect specimens," "to map soil types and forest cover," or "to measure water parameters."

- *Methods.* Describe the field techniques that you used, such as electrofishing, the instrumentation that you used, and any modifications you made to field methods or instruments to suit conditions as you found them.

- *Field trip study area.* Make at least a sketch map to orient your reader. In most cases, more serious cartography is required. Use a base-map series, such as NTS or OBM, and add your own data sets relevant to your project. In the text of this section, describe the general location and topography of the area where you conducted the field study.

- *Background data.* Include the following information about the fieldwork: personnel in the crew; places, dates, and times; and the equipment and transportation required. When you have to brief or instruct a field crew, such as a tree planting crew, include the preparation in this section. Include also any previous research on the scientific or technical subject that you are investigating in the field. Indicate the general weather conditions during the fieldwork, since these can affect the efficiency of the work and the data itself.

Organize your field activities in the report body. Keep good notes while in the field; they are the source of data for your report. Group similar activities or information under descriptive headings (Geomorphology in one section, Geophysical Data in another), and use topic headings, not complete sentences or questions.

Avoid using chronological order for your report wherever possible. The sequence of when events happened is generally less important than the types of activities undertaken and the types of results from the investigations. For example, a magnetometer survey yields data best presented to your reader on a map of rock formations and then described and interpreted in your text; the actual process of running grid lines with the magnetometer is not important to the reader.

Describe activities thoroughly and accurately. Choose only activities useful to your audience and purpose. Depict field information in graphic form, and leave analysis of data to the end of the report. Discuss only those events that were relevant to the purpose of the trip—field trips are fun, but avoid personal anecdotes in your report.

The style of reporting in a field trip report should define scientific or technical terms. Be concise and edit your text carefully, eliminating unnecessary detail and vague generalizations.

Concluding the Field Trip Report

Include the following in your conclusion.

- *Summary.* Review your most important activities. Include the most significant results.
- *Results.* Analyze your data. Use tables and figures. Evaluate the success of your trip. Compare your reported results with the trip's goals.
- *Appendix.* Attach at least your field notes; these are a necessary check against the accuracy of the data presented in the body of the report.

RESEARCH REPORTS

We answer questions about natural processes by experimentation and by looking in scientific literature for reports of experiments that answer similar research questions. When you write up the results of your own experimental work, you are contributing to that body of scientific and technical literature. Your report will review the experiments of others and describe your own experiment or investigation.

The research report describes mostly original work, called *primary research.* A report section called the *literature review*, usually presented as part of the introduction, is the only secondary research found in this type of report. This section summarizes published reports dealing with similar investigations and information. The majority of the report text describes the purpose, methods, and results of your own scientific or technical research. It concludes with a discussion of the experimental results: a preliminary analysis, a summary of important details, and comparison of results with research goals.

Experts write research reports for other experts. When you read the examples of report format and organization in this book, keep in this mind. Their content may not interest you because it is not in your field of study, but their format is universal and useful. However, readers of scientific and technical literature share specific interests and expertise with the writer. Therefore, the goal of writing a research report is to inform those expert readers, not to interest, educate, or entertain them.

Parts of the Research Report

Add to your report the elements necessary to guide your reader and to meet the demands of the situation: title page, table of contents, illustrations list, abstract, reference list, and appendix.

Divide the body of your report into the sections listed and described below. Note that you can also use this sequence of topics for articles describing your research and written for publication in scientific and technical journals. In the latter case, follow the format guidelines provided by the journal and its editorial staff.

- *Introduction.* Describe the general subject area of the investigation, the specific research goals, and the scope of the research, both as planned and as experienced. You can state your research goals as a hypothesis, as a question to be answered, or as a simple statement of what was studied.

- *Literature review.* Include a literature review that describes reports of previous research designed to answer similar questions. In college, write an extensive review to demonstrate a comprehensive knowledge of published literature in your research field. Make it a separate section with its own heading. In the workplace, describe only papers that helped you to limit and focus your research project or to identify effective methods or instrumentation. Include the review as a section within the introduction to your research paper.

- *Materials and methods.* In this section, describe the design of your research project, including relevant scientific and/or technical procedures, and the materials, instrumentation, and other facilities you used in the project. Your goal is to describe the experiment clearly enough that your reader could set up and run the same experiment with similar results.

- *Results.* This is the most important section in your report, answering the questions in your statement of purpose. Present the raw data from your observations and your measurements in the form of tables and figures, and point out the significant features and trends in the data.

- *Discussion.* Provide some interpretation of your results. Evaluate the results in terms of whether the investigation met the research objectives, any problems with the methods or results, inferences that you can draw from the data, and implications for further research in the same area.

PROPOSALS

The purpose of writing a proposal is to describe a proposed plan of work that will solve a problem or improve a process and to get the reader to finance that proposed plan of work. North American corporations and governments routinely contract private-sector work in this manner. A second important purpose in writing a proposal is to get you and your company accepted to do the work. A good plan of work that solves scientific and technical problems is essential to a good proposal, but a good plan can be offered to other contractors. Therefore, your proposal must also convince the reader that you and your company are best qualified to do the proposed work.

Types of Proposals

Internal proposals are submitted within your company to supervisory staff. They are suggestions for new or improved processes and facilities. Their format is usually brief and informal. *External proposals* are submitted to clients outside your company and result in contracts that make money. Their format is longer, more detailed, and more formal in style.

Unsolicited proposals are submitted without a prior request for a proposal. Researching the market and networking with associates will identify potential clients who would read an unsolicited proposal. Informal meetings with such clients will show you their needs and how to address them in the proposal.

Solicited proposals are submitted in response to a request from a potential client, often a government agency. These requests usually appear in two forms. An Information for Bid (IFB) is a request for standard products and services, such as cleaning or food services. A Request for Proposal (RFP) is a request for customized products or services; most scientific or technical services fall into this category, such as the engineering of structures or geological field surveys.

Keys to Writing Successful Proposals

1. *Understand the client's problem or situation clearly.* This is the single most important factor in your proposal. Your understanding of the client's needs must be thorough and accurate; do your homework on the client's situation, then write this understanding into well-structured and well-expressed sections of your proposal. The client's recognition that you have understood his or her situation or problem is the key motivator in giving the contract to you and your company.

2. *Write a practical, detailed plan of work.* The procedures you describe must be easy to achieve; usually, this means standard processes with some modification to suit the client's needs. Provide enough detail to show your reader that you are familiar with the type of work being described.

3. *Estimate a fair price for the work.* The total value of the work need not be the lowest price possible. The client will weigh the costs of submitted proposals against other factors such as the design of the work and the reliability of the company itself. However, your price must be competitive.

4. *Provide your company's credentials.* If all the other factors are equal among competing companies, the contract will usually go to the company with the most experience. Therefore, include a section describing your company's previous projects and the qualifications of the principal staff members who will be responsible for the proposed work.

Proposals require more than a good technical description; they require persuasive writing. This means including sections that show how the future benefits of the project outweigh the costs. These sections will follow the basic plan that persuades: a detailed understanding of the problem or opportunity, detailed planning, and company credentials that qualify you to do the work.

Guidelines for Proposals

For a solicited proposal, follow IFB or RFP guidelines exactly. These are usually published in some detail, including the nature and extent of the work along with format requirements for the proposal document itself. Look for key language in the proposal. Frequently repeated scientific or technical words and phrases are clues to the essential concerns of the potential client: address these concerns in detail in your proposal. When revising and editing the proposal, ensure that these key words and phrases appear frequently. Such phrases found in government RFPs often represent the political agenda of the government in power and reflect the areas where they are prepared to fund projects that meet their criteria.

Elements of a Proposal

The following is a format for an unsolicited proposal. Since projects vary considerably, you will not need to cover all of the following topics in equal detail. Be sure to consider each one carefully before deciding what to write. Then add, combine, re-organize, or omit topics to suit your proposal. Above all, research the project thoroughly. Good proposals are built on complete, accurate detail.

Summary

Include a summary in either the front matter or the body of your report. Include in your summary all the major points in your proposal: the definition of the problem, the proposed program of work, your company's qualifications and experience, a timeline for work, and the budget estimates.

Introduction

Outline your subject and purpose, and describe what kind of work you propose and what its outcome will be. Describe the problem faced by the potential client as you understand it, using short, detailed paragraphs. Quantify each aspect of the problem wherever possible, and check your lab or field-test measurements, as well as secondary source data, for accuracy.

Describe the current technologies that you intend to apply to the problem. Most readers of proposals are well educated but unfamiliar with specific technologies used in other fields. For example, describe the theory of operation of state-of-the-art equipment and support it with a simple diagram or flow chart.

Work Plan

Organize your proposed work plan into sections, with subheadings identifying the major tasks to be accomplished. This makes it easier for the reader to absorb details and understand them. Be specific and detailed in your description of the steps in each task. To do this, refer to your research notes, which must be thorough, covering all aspects of the problem. They may include your own estimates and test measurements, as well as secondary research in reliable, current scientific and technical literature.

Check your facts and figures; mistakes in measurements undermine client confidence. Revise your text carefully for errors of expression when you are explaining principles and processes. Also, try to avoid generalities. It is tempting in a proposal to shorten the work of writing by estimating quantities of materials or time required to complete the proposed work without the benefit of extensive experience or research. The clue to generality is the relative term. A statement like, "This process will take only a short time" will signal to a proposal reader that you do not have an estimate of the time required, since "short" is a relative term. In a proposal, you must commit to specifics wherever possible.

Avoid statements assuring the reader of your enthusiasm and competence. These mean little on paper, since anyone can say them without meaning them. What speaks most clearly to the reader about your enthusiasm and ability to do the proposed work is the work that you have already put into the proposal, both the research and the writing. An accurate, detailed, well-organized plan of work will show the reader that you can—and want to—do the proposed work.

Include appropriate illustrations to support the principles and processes in your proposed plan of work. Tables, diagrams, flow charts, cutaway views, photographs, and instrument layouts are some of the ways in which you can show the reader how the work would be done. Include also any estimates of cost that you have received from potential suppliers or subcontractors.

Qualifications

Describe how you and your company are qualified to do the work in the proposed plan. Identify each staff member who would work on the proposed project, and include a résumé for the designated supervisor and a paragraph for each of the others.

List the major projects that your company has completed, as potential clients will want to see what you have done for other companies or agencies. Provide brief descriptions of projects similar to the one in your proposal, illustrated with photographs. List also the equipment and facilities for the proposed project that the company owns.

Describe how your company is structured, including chain of command, local offices, facilities, and corporate partnerships. Build a picture in this section of a company ready and able to do the work. Often it is experience that wins; if the plans of work and the costs among competing bids are roughly the same, the contract will go to the company with the most experience.

Scope and Methodology

Describe precisely and realistically how much work you plan to do. Most readers of proposals will be concerned about potential time and cost overruns. Therefore, set reasonable limits on the proposed work.

Describe also how you will do the work, and an overview of the tasks and steps necessary to achieve the goal. Explain briefly any scientific or technical procedures required to complete each task, keeping in mind that the client reading the proposal may not be familiar with your company's technologies.

Facilities, Personnel, and Duration

Use a bulleted list to set out the equipment and estimates of supplies required to complete the project. Include any equipment rentals or contracted services necessary.

List the names and positions of the company personnel who will work on the project, and include the names of contracted persons or companies required to complete the work.

Estimate the total time required for the project, then provide a detailed schedule of tasks and procedures. If possible, present this information in the form of a chart or table. Remember to be accurate and complete in your estimates; if you underestimate or overestimate the time required for the project, and this schedule becomes part of the contract, the time factor will become a problem for both parties.

Budget

Set out the budget for the project in the form of a financial statement with standard line items and associated costs. Divide your project budget into direct and indirect costs.

Direct costs include salaries and benefits, travel costs, equipment purchases and rentals, contracted services, materials, and supplies. Research each of these thoroughly to provide accurate estimates.

Indirect costs, also called overhead, are the many small costs of doing business. These include clerical services, utilities for offices and buildings, and maintenance of vehicles and facilities. Since these indirect costs are too numerous and too small to estimate individually, express them as a percentage of direct costs, usually about 10 percent for large projects up to 30 percent for smaller ones.

Reports and Benefits

Most projects of any length and complexity require the project manager to submit regular progress reports to his or her supervisor, and to the client. In this section, indicate the personnel responsible for writing progress reports, specify the reporting intervals, and determine the form that the progress reports will take.

Most important, end with a section called "Benefits" that details the advantages of the project to the proposed user. Begin by describing the outcomes of the proposed plan of work, or what the client will have when you are finished. Then identify the benefits to the client in terms of the problem or situation you described in the opening sections. Organize your list of benefits in more- to less-important order. By placing the benefits section at the end of your proposal, you will leave your reader with a positive impression of the proposed plan of work and of your company.

Appendix

An appendix is not required but may be useful in providing a depth of detailed information that would not be appropriate in the body of the report. It can also provide further testimonials to the company's credentials, since this is a key element in any successful proposal. Items that can prove useful are testimonial letters or e-mails from satisfied clients, detailed task schedules presented as Gantt or milestone charts, and evaluations of previous projects from third parties, both qualitative and quantitative.

Preparing a good proposal takes time for research and writing. In the end, your proposal may not win the contract, but a good proposal will always leave a positive impression and lead to better opportunities in the future for you and your company.

FEASIBILITY REPORTS

A *feasibility report* presents the results of a feasibility study, a series of investigations into whether a proposed plan of work is viable. Such studies may consider a broad range of factors, or only one or two: engineering and technical possibilities, economic practicalities, ecological balance, social values, or psychological health. Feasibility studies that deal with large and complex issues involving science and technology usually assess the feasibility of two or more options for solving the central issue or problem.

Design the feasibility study itself to answer a clear and focussed question—for example, "The purpose of this investigation is to recommend to the Forest Pest Management Institute the most appropriate method of forest pest control in the watershed of the Nashwaak River in New Brunswick." Investigating something out of scientific or technical curiosity is not a feasibility study. Your purpose is to recommend something specific to someone specific.

Feasibility studies are often carried out in teams of experts, and thus the teams often write the reports that describe these studies. A good example is the environmental impact study. When a new highway is proposed, teams of civil engineers, fisheries and wildlife biologists, environmental technologists, geotechnical technologists, urban planners, and archaeologists study alternative routes for their environmental impact. Each contributes to the final feasibility report that recommends one proposed route over the other possibilities.

Parts of a Feasibility Report

- *Introduction.* Begin with the purpose and scope of the feasibility study. Whatever the original goals, the amount of information collected during the course of the investigation will depend on time and resources. It is important to reveal to your reader how much data you were able to collect in order to answer the feasibility question stated in the purpose. Describe briefly the problem that you investigated and the methods you used to acquire information.

- *Discussion.* In this section, present the results of your investigations, organized into a logical sequence of subsections under descriptive headings. Illustrate your data sets with appropriate types of graphic presentations: graphs, charts, photos, maps, histograms, and so on. Use the illustrations to interpret the results and draw appropriate inferences from the data.

- *Conclusions.* Summarize the data and review the purpose of the study. A vertical list is often the best way to present the most significant findings of a feasibility study.

- *Recommendations.* Make a bulleted list of recommended actions, basing each one on the conclusions developed directly from the data reported from your investigations. Write in the imperative mood: "Apply BT in 300:1 concentrations to the spruce–jack pine component of the upper watershed forest stands." Remember that recommendations are your professional judgments; they are for your reader to consider and act on at his or her discretion.

GOVERNMENT REPORTS

Scientists and technologists working for federal, provincial, or municipal governments in Canada will write reports explaining scientific and technical principles and processes and applying them to current public problems and issues. The list below presents an outline of topics for a typical government report.

Parts of a Government Report

- *Introduction.* Include the official terms of reference for your study as set out by the authorizing government body, such as a planning council or policy review committee. Also include a review of the problem that you investigated, the public issue or issues that generated the research, and current approaches by other governments. Describe in the background section any history of the particular issue or problem that the report addresses.

- *Report body.* Address one or more of the following issues:
    - Roles of public bodies
    - Directions for improvement

    - Concepts, principles, information
    - Feasibility
    - Legal questions
    - Public impact
    - Policy and implementation issues
    - Planning and resource issues
    - By-law and statute conflicts and issues
    - Evaluation processes

- *Conclusion.* Finish the report with a summary of findings and recommendations if they are required by your terms of reference.

- *Front and back matter.* A government report is usually a formal report, requiring all major elements. Front matter will usually include a title page, letter of transmittal, table of contents, and an executive summary instead of an abstract. If the report has graphics, include a list of illustrations. The back matter will include a reference list, since you will refer in the report to scientific and technical literature and to Canadian statutes, and an appendix, since you will have supplemental material to support your text.

CHECKLIST FOR TYPES OF REPORTS

Be sure that you have answered the following questions:

Instructions and Manuals

- Did you make research notes while doing the procedure?
- Did you keep it simple? Know the procedure? Put yourself in the reader's place?
- Did you choose good illustrations, and test your instructions on readers?
- Did you include an overview and materials list for each task?
- Were safety messages required? Are they properly formatted?
- Are instructions grouped under headings, written in the imperative mood, and fully illustrated?
- Did you conclude with maintenance? Troubleshooting? Testing?

- Is the manual divided into chapters? Does it have detailed overviews for audiences, and appendix reference material?

Progress Reports

- Who is your reader? What is the progress-report schedule?
- Is the format consistent for all project reports?
- Did you gather operational and planning information?
- Is the content organized into time, tasks, and topics?
- Did you answer management's key questions (is it on time, is it on budget, and are there any problems)?

Field Trip Reports

- Did you include research notes from the field study? Data recorder? Journal?
- Does the introduction state aims and purpose and include a map of the study area and background data?
- Is content organized into similar activities or types of data, under descriptive headings?
- Are results presented with graphics? Are data analyzed with standard methodology?
- Does the appendix include field notes?

Research Reports

- Does the introduction describe research aims, goals?
- Does the literature review section summarize and evaluate relevant published research?
- Does the materials and methods section include research design (materials, instrumentation, facilities)?
- In the results section, are data presented in tables and figures and described and interpreted in the text? Did you double-check the data for accuracy, and edit carefully?
- In the discussion section, is there an evaluation of the research plan? Inferences from data? Implications for further research?

Proposals

- Was the proposal solicited or unsolicited? A response to an IFB or RFP?
- Are the client's needs described clearly? Is there a practical, detailed plan of work, including a good price and company credentials?
- Does the introduction include a work plan? Are client concerns described completely and accurately? Did you check the date? Describe technologies?
- Is the work plan organized in a logical sequence, with accurate information and a focus on specific estimates?
- Are relevant qualifications, including previous projects, described for the company and individual personnel?
- Did you address practical limits, how much work is involved, and how work will be achieved?
- Did you list all equipment, supplies, personnel required, and a timeline of main tasks using accurate estimates?
- Does the budget include accurate estimates of direct and indirect costs?
- Does the proposal include a progress-report schedule?
- Did you describe outcomes of the proposed plan and benefits to the client, organized from more to less important?
- Is an appendix required? Testimonials? Detailed timelines? Details of company projects?

Feasibility Reports

- Is there a clear statement of purpose of the feasibility study?
- Is there good team communication? A clear division of writing jobs?
- Does the introduction state the purpose, scope of investigations, and methods used?
- Is the discussion organized under descriptive headings? Is there an interpretation of main points? Does it present data using illustrations?

- Does the conclusions section include a summary, noting significant findings?
- Is there a bulleted list of recommended actions?

Government Reports

- Did you adhere to the official terms of reference in the introduction, and cover the background of the issue?
- Are all issues addressed in the body, and organized into logical sequence with descriptive headings?
- Are all required elements included in the front and back matter? Are they in the correct order?

Chapter 5

Report Documentation

Overview: This chapter sets out the correct format for documenting published sources of information used in scientific and technical reports.

CHOOSING YOUR SYSTEM

Your first decision is to choose the system of documentation for your report that is appropriate to your situation. No single system of documentation is universal. Documentation in report writing often follows the practices of professional journals that publish research articles, styles based on one of the major systems but modified by journal editors to meet the needs of their specific publications and readers. Similarly, workplace report writers often modify their documentation styles to meet the needs of specific companies, government agencies, or individual writers and editors.

Most scientific and technical writers document their sources using one of three systems: the Council of Science Editors (CSE), the American Psychological Association (APA), and the Modern Language Association (MLA). In general, one uses CSE in the pure and applied sciences, APA in the social sciences, and MLA in the humanities. An alternative system, widely used in the humanities, is the Chicago Manual of Style. If you are a college student, refer to the system required in specific courses at your college or university. Consult your professors for help with CSE, APA, or MLA documentation; they know which documentation style they require in their courses. If you are in the workplace, consult other reports produced by professionals in the same department, company, or government agency to determine the accepted style of documentation.

Questions often arise about the details of how to cite or reference specific kinds of published information. Fortunately, answers for documentation questions are readily available. The CSE, the APA, and the MLA each publish a comprehensive guide to their system of documentation. These guides show example citations and reference entries for most types of formal and informal information sources used by writers.

Internet search engines regularly list pages from North American college and university Web sites explaining the popular documentation systems and giving detailed examples of references and citations. Try the following:

- The Engineering Communication Centre at the University of Toronto.
- The New Guide to Writing Research Papers at Monroe Community College
- The OWL (Online Writing Lab) at Purdue University
- The Writing Center at the University of Wisconsin–Madison

Scientific journals such as the *Journal of Soil and Water Conservation* often use variations on the major systems. Consult these specific sources if you are submitting scientific or technical articles to a journal for publication.

THE IMPORTANCE OF CITING AND LISTING SOURCES

You must cite your sources in the text of your report every time you use them. You must also list complete information about each of your sources at the end of the report text. This list is a separate element; call it "References," "List of References," "Bibliography," "Literature Cited," or "Works Cited" depending on the system of documentation you are using. Determine the correct name for the reference list before creating the references page in your report.

Your citations and reference list go together to give your reader complete information about not only which parts of your report came from secondary sources but also how to find those secondary sources. Your citation follows a graphic or text passage showing the secondary source of information, and it points to an entry on your reference list giving complete information about the source so that your reader can look it up.

Cite your sources thoroughly; partial citing of sources is a common failing in student reports. Place a citation in your report the first time you use source information and every time thereafter. Cite the sources of information you used to write your text, and cite the sources of tables and figures used to illustrate your text.

Frequent citations are both necessary and desirable in scientific and technical reporting. Students often hesitate to place several citations on a report page, yet this is exactly what the report writer is supposed to do: consult good secondary sources and give the information to the reader. Citations show the reader that you have understood more than how to carry out your own field and lab procedures. In fact, many college report assignments specifically require students to summarize available scientific and technical literature.

Extensive secondary research can be essential to the good design of your own research before you begin work. Published scientific literature that describes work similar to yours will help you to interpret your own results and must appear in the literature review section of a research report. Summarizing current information from published sources in your report also accomplishes the following specific purposes:

- provides a general background of research similar to your own field or lab studies presented in the report
- provides a model or method for your own field or lab research presented in the report
- provides an authority for analysis or conclusions in your report
- provides a more complete explanation or set of data on your report topic

Using Citations

Place a citation at the end of each borrowed section in your report. Borrowed sections are paraphrased summaries of the original source material. Borrowed sections vary in length: they may be a sentence, a paragraph, or a short section of two or three paragraphs under a heading. Include the citation at the end of each completed section of text before you begin to write the next one.

Borrowed sections of a report may combine two or more sources into a paraphrased passage. In this case, add two or more citations after the text passage. Borrowed sections may combine two or more different passages from the same source, such as a textbook. In this case, add two or more citations that include the different page numbers for these source passages.

You may refer to the author of your source directly in your text, as in the following example: "Schneider (2004) showed isolation to be an effective method of vibration control." This technique effectively makes your reader aware of your sources of information. The author becomes part of the sentence, fully integrated with the report.

Plagiarism

Plagiarism is the theft of the work of others either knowingly or unknowingly by improper or inadequate documentation. At most North American colleges and universities, plagiarism is a breach of academic integrity. The penalties for such conduct are set out in college academic regulations and published in campus documents. The penalties in the workplace can include lawsuits and loss of employment. Handing in a report with your name on it implies that its contents are your original work, unless otherwise indicated through the documentation system. To avoid plagiarism, write a citation for each text passage or graphic illustration from a published source, identifying the source of information for your reader. The appearance of citations in your text clearly separates your own data from the work of others.

Citations are short and give only a small part of the information that your reader needs to locate the published works that you have summarized in your report. Their purpose, then, is to point to a published source on your reference list. The reference list gives complete available information for each published source that you consulted and included in your report text. In this way, the citations and the reference list work together to help the reader identify a source of information without getting in the way of reading and understanding the report.

COUNCIL OF SCIENCE EDITORS (CSE)

The Council of Science Editors (CSE) was originally called the Council of Biology Editors, so you may hear this format referred to as either CSE or CBE style. The 6th edition of *Scientific Style and Format: The CBE Manual for Authors, Editors, and Publishers* describes two systems of documentation: the name-year system and the citation-sequence system.

The name-year system links the name of the author with the year of publication. The name-year reference list arranges your sources alphabetically by author surname.

The citation-sequence system uses simple numerals in sequence as citations in the text of your report. The citation-sequence reference list arranges your sources numerically in the order you first cite them in your text.

CSE Name-Year System

CSE name-year citations place the name of the author with the year of publication together in parentheses, or round brackets, like this: (Smith, 2004).

Forms for CSE Name-Year Citations

- *One Author.* Separate author and date with a comma: (Smith, 2004).
- *Two Authors.* Put them both in the citation: (Smith and Jones, 2004).
- *Three* or *More Authors.* Write only the first name followed by the abbreviation *et al.*, Latin for "and the others": (Smith *et al.*, 2004). Do not use *et al.* in your reference list entries.
- *No Author.* Show the author as anonymous, enclosed in square brackets: [Anonymous].
- *Source with More Than 10 Pages.* Include a page reference in the citation after a colon: (Smith *et al.*, 2004: 245). (See Box 5.1.)

CSE Name-Year Reference List

Begin the CSE name-year reference list on a separate page at the end of your report; it will be the last numbered page. Give it the title REFERENCES or REFERENCES CITED. Use bold capital letters for the heading and centre it at the top of the page. Include only sources of information that you have cited—that is, written citations for—in the text of your report. Box 5.2 shows an example of a CSE name-year reference list.

Forms for CSE Name-Year Reference Lists

Format your reference entries using single spacing and hanging indentation, which indents the second and successive lines of the entry. See Appendix B for a tutorial showing how to format a reference list using Microsoft Word XP.

Arrange entries in a name-year reference list alphabetically by author surname. Give the correct information in the correct order for each of your sources, following the conventions for reference entries shown below. Arrange items within entries in descending order of importance, beginning with the author, date, and title. Note that the punctuation of items within the entry is mostly with periods. Separate units within the item with commas, and use a colon in two cases: to introduce the second part of a two-part title, and to separate the place of publication from the name of the publisher.

The following examples show how to arrange CBE name-year reference list entries for commonly used types of sources.

11

The most important monument at Copán is Altar Q. This limestone slab carries glyphs on all four sides and on the top. It provides us with some details about the earliest events in the reign of the first king, Kinich Yax Kuk Mo, as well as the dynastic succession of all the rulers of Copán. In one inscription, Kinich Yax Kuk Mo is said to celebrate the great, period-ending date in the Mayan calendar, 9.0.0.0.0, or December 11, 435 A.D. In all, Altar Q carries information concerning sixteen rulers of the city. The last date at Copán is associated with the last ruler, Yax Pasah. This is July 24th, 805 A.D., after which the record is silent (Coe, 1975: 142; Stuart, 2002).

Box 5.1: CSE Source with More Than 10 Pages

14

REFERENCES CITED

Coe MD. 1993. From huaquero to connoisseur: the early market in pre-Columbian art. In Collecting the pre-Columbian past. EH Boone, editor. Washington, DC: Dunbarton Oaks. pp. 271–90.

GB Online. 2002. Maya codices. GB Online's Mesoamerica. [Online]. <http://pages.prodigy.com/GBonline/mesowelc.html>. Accessed 2004 Mar 20.

Hammond N. 1982. The Maya. New Jersey: Rutgers Univ Press.

Box 5.2: CSE Name-Year Reference List

- *Single Author*

Erjavec J. 2000. Automotive technology: a systems approach. 3rd ed. New York: Delmar Thomson Learning. 1343 p.

- *Two Authors*

McNair HM, Miller JM. 1997. Basic gas chromatography: techniques in analytical chemistry. New York: Wiley-Interscience. 224 p.

Note that the initials of the second author appear after the author's surname.

- *Multiple Authors*

Griffiths AJF, Miller JH, Suzuki DT, Lewontin RC, Gelbart WM. 1996. An introduction to genetic analysis. 6th ed. New York: WH Freeman. 916 p.

- *Two Publications by the Same Author*

List entries in chronological order by date of publication, earliest first:

Snyder JP. 1987. Map projections: A working manual. US Geological Survey Professional Paper 1395. Washington, DC: USGPO.

Snyder JP. 1993. Flattening the earth: two thousand years of map projections. Chicago: Univ of Chicago Press. 365 p.

- *Two Publications by the Same Author in the Same Year*

List entries in alphabetical order by publication title. Then assign lowercase letters to each date of publication in order (2002a, 2002b, etc.).

Long JR. 2000a. A 5.1–5.8 GHz low-power image-reject downconverter. IEEE J Solid-State Circuits 35: 1320–1328.

Long JR. 2000b. Monolithic transformers for silicon RF IC design. IEEE J Solid-State Circuits 35: 1368–1382.

- *Book with Editor(s)*

Dorfman M, Thayer RH, editors. 1996. Software engineering. Toronto: J Wiley & Sons Canada. 546 p.

- *Book Published in New Edition*

Diamond WJ. 2001. Practical experiment designs for engineers and scientists. 3rd ed. Toronto: J Wiley & Sons Canada.

- *Book Published in Volumes*

Verschueren K. 2001. Handbook of environmental data on organic chemicals. 4th ed. 2 volumes. Weimar (TX): Culinary and Hospitality Industry Publications Services.

- *Article in an Encyclopaedia*

Hinrichs T. 1992. Geothermal power. McGraw-Hill Encyclopedia of Science and Technology. 7th ed. New York: McGraw-Hill 8: 83–87.

- *Article in a Scholarly Journal*

Brett NC. 1991. Language laws and collective rights. Canadian J Law Jurisprudence 4(2): 347–360.

- *Article in a Monthly Periodical*

Ricciardelli A, Pizzimenti D, Mattei M. 2003. Passive and active mass damper control of the response of tall buildings to wind gustiness. Engineering Structures 25(9): 1199–1209.

- *Article in a Newspaper*

Palmer K. 2003 May 7. West Nile plan outlined: province vows to increase spending. Toronto Star; Sect B: 5.

- *Article without an Author*

[Anonymous]. 2003 July 11. Biotech crops get third-world boost. Globe and Mail; Sect C: 4.

- *Video Recording*

Decision making and problem solving. [Videocassette]. Coast Community College District. Toronto: TVOntario; 1990. 28 min, sound, colour, ½ in.

- *Article in an Online Journal*

Wang F, Juniper SK, Pelegrí SP, Macko SA. Denitrification in sediments of the Laurentian Trough, St. Lawrence Estuary, Quebec, Canada. Estuarine, Coastal and Shelf Science. [Serial Online]. 2003; 57 (3): 515–522. <http://www.sciencedirect.com/science/journal/02727714>. Accessed 2003 Jul 16.

- *CD-ROM*

Nichols M. 1995. High-tech artificial limbs. Maclean's. March 13, 1995. The Canadian Encyclopedia 2001. [CD-ROM] Toronto: McClelland & Stewart; 2000.

- *E-mail*

Birnstihl J. Re: Update. [Personal E-mail]. info@cdta.bidcon.net. Accessed 2003 May 12.

- *Web Page*

Maddison DR, Maddison WP, Schulz KS, Wheeler T, Frumkin J. 2001. The Tree of Life Web Project. [Online] <http://tolweb.org>. Accessed 2003 Jul 15.

- *Article from a Subscription Service or Online Database*

Bergman, B. 2003 June 23. Born to be high and wild. Maclean's 116 (25). EBSCOhost Academic Search Premier. Rogers Media, Publishing Ltd. Item 10046413.

CSE Citation-Sequence System

CSE citation-sequence citations are simply Arabic numerals placed in parentheses, for example: (1). Like name-year citations, place citation numerals at the end of each borrowed section in your report. Borrowed sections may combine two or more sources into a paraphrased passage with two or more citation-sequence numerals after the text passage.

Assign a numeral to each of your sources in the order that you first use and cite its information in your report. For example, if Smith is the source of your first borrowed section of text, give it the numeral "1." This number then identifies Smith in your reference list. If Jones is the next source you have used in your report, assign it the numeral "2," and so on. Notice that each source has only one citation numeral. If you decide to use material from Smith later in your report, cite Smith with the numeral "1" again. Do not assign Smith another numeral. (See Box 5.3.)

CSE Citation-Sequence Reference List

Place the CSE citation-sequence reference list on a separate page at the end of your report. It will be the last numbered page. Give it the title REFERENCES or REFERENCES CITED. Use bold capital letters for the heading centred at the top of the page.

Include only sources of information that you have cited—that is, written citations for—in the text of your report. Box 5.4 shows an example of a CSE citation-sequence reference list.

Format your reference entries using single spacing. Use the hanging indentation format for a numbered list. The second and successive lines of each entry should begin directly below the first letter of the first word on the first line of the entry. Arrange your reference entries in numerical order according to the numbers that you have assigned to your sources in the report text.

Forms for CSE Citation-Sequence Reference Lists

Citation-sequence reference entries contain the same information in the same order as those for name-year reference entries (see the examples of name-year reference entries on page 121 for type, order, and format of information). Only the order of entries in the list is different. List the citation-sequence sources in numerical order: each has a number that you assigned to the source when you first paraphrased its information in your report.

The format for the citation-sequence list entries is the indented block with the number of the entry on the left margin of the page, as shown in the sample citation-sequence reference list (Box 5.4).

10

The most important monument at Copán is Altar Q. This limestone slab carries glyphs on all four sides and on the top. It provides us with some details about the earliest events in the reign of the first king, Kinich Yax Kuk Mo, as well as the dynastic succession of all the rulers of Copán. In one inscription, Kinich Yax Kuk Mo is said to celebrate the great, period-ending date in the Mayan calendar, 9.0.0.0.0, or December 11, 435 A.D. In all, Altar Q carries information concerning sixteen rulers of the city. The last date at Copán is associated with the last ruler, Yax Pasah. This is July 24th, 805 A.D., after which the record is silent (2).

Box 5.3: CSE Citation-Sequence Citation

14

REFERENCES

1. GB Online. 2002. Maya codices. GB Online's Mesoamerica. [Online]. <http://pages.prodigy.com/GBonline/mesowelc.html>. Accessed 20 Mar 2004.

2. Coe MD. 1993. From huaquero to connoisseur: the early market in pre-Columbian art. In Collecting the pre-Columbian past. EH Boone, editor. Washington, DC: Dunbarton Oaks. pp. 271–90.

3. Hammond N. 1982. The Maya. New Jersey: Rutgers Univ Press.

Box 5.4: CSE Citation-Sequence Reference List

AMERICAN PSYCHOLOGICAL ASSOCIATION (APA)

The APA system has been in use since 1929. The 5th edition of the *Publication Manual of the American Psychological Association* was released in 2001. In APA style, citations consist of the author's surname and the date of publication in parentheses (round brackets). Complete information about each source is listed alphabetically by author surname at the end of the report in the reference list.

APA Author-Date Citations

To make an APA citation for a paraphrased section in your report, place the name of the author with the year of publication together in parentheses, like this: (Smith, 2004) (see Box 5.5). If you quote from your source directly in your report, add the page number to your citation: (Smith, 2004, p. 105).

Forms for APA Author-Date Citations

Introduce an APA author-date citation in the text with a phrase that includes the author's surname, followed by the date in parentheses: "When similar metal fibres were tested by Johnson and Hubbard (2002), tensile strengths of 600 to 800 g were recorded."

- *One Author.* (Smith, 2004)

- *Two Authors.* (Johnson & Hubbard, 2002). Note the use of the ampersand (&).

- *No Author.* Cite the source by using the first two or three significant words of the title, like this: ("Biotech crops," 2003).

- *Three or More Authors.* List all authors the first time you cite their work, like this: (James, Harrison, Wilson, McAlpine, & Fingaard, 2003). The second and successive times you cite their work, list only the first name followed by "et al.," like this: (James et al., 2003).

10

The most important monument at Copán is Altar Q (Coe, 1975). This limestone slab carries glyphs on all four sides and on the top. It provides us with some details about the earliest events in the reign of the first king, Kinich Yax Kuk Mo, as well as the dynastic succession of all the rulers of Copán. In one inscription, Kinich Yax Kuk Mo is said to celebrate the great, period-ending date in the Mayan calendar, 9.0.0.0.0, or December 11, 435 A.D. In all, Altar Q carries information concerning sixteen rulers of the city. As epigrapher David Stuart (2002, p. 16) has concluded, "The last date at Copán is associated with the last ruler, Yax Pasah. This is July 24th, 805 A.D., after which the record is silent."

Box 5.5: APA Citation Style

- *Corporate Author.* Write out the name of the organization the first time you cite its work, and include the abbreviation of that organization in square brackets: (The American Radio Relay League [ARRL], 1997). Use only the abbreviation in second and successive citations: (ARRL, 1997).

- *E-mail.* (J. Birnstihl, personal communication, May 12, 2003). Note that APA documentation does not include personal communications in the reference list.

Forms for APA Author-Date Reference Lists

List all sources you cited in the text of your report under the centred heading References (see Box 5.6). It is acceptable in APA style to format entries in reference lists using either hanging indentation or regular indented paragraph format. Whichever format you choose, use it consistently throughout the reference list. Double-space your entries. List your sources in alphabetical order by author's surname. If you have more than one source by the same author, list the sources by date of publication, oldest first. If you have more than one source by the same author in the same year, list the sources in alphabetical order by title, and add lower-case letters to the dates of publication, like this:

Smith, K. L. (2003a). *Book A* ...

Smith, K. L. (2003b). *Book B* ...

Titles of separate works such as books and journals are italicized or underlined. The dates of publication following the author's name are placed in parentheses.

Examples of APA Author-Date Reference Entries

- *Single Author*

  Erjavec, J. (2000). *Automotive technology: A systems approach* (3rd ed.). New York: Delmar Thomson Learning.

- *Two Authors*

  McNair, H. M. and Miller, J. M. (1997). *Basic gas chromatography: Techniques in analytical chemistry.* New York: Wiley-Interscience.

14

REFERENCES

Coe, M. D. (1993). From huaquero to connoisseur: The early market in pre-Columbian art. In E.H. Boone (Ed.)., *Collecting the pre-Columbian past* (pp. 271–90). Washington, D.C.: Dunbarton Oaks.

GB Online. (2002). Maya codices. GB Online's Mesoamerica. Retrieved March 20, 2004, from http://pages.prodigy.com/GBonline/mesowelc.html

Hammond, N. (1982). *The Maya*. New Jersey: Rutgers University Press.

Box 5.6: APA Author-Date Reference List

- *Multiple Authors*

  Griffiths, A. J. F., Miller, J. H., Suzuki, D. T., Lewontin, R. C., & Gelbart, W. M. (1996). *An introduction to genetic analysis* (6th ed.). New York: W. H. Freeman.

- *Two Publications by the Same Author*

  List entries in chronological order by date of publication, earliest first.

  Snyder, J. P. (1987). *Map projections: A working manual.* U.S. Geological Survey Professional Paper 1395. Washington, DC: USGPO.

  Snyder, J. P. (1993). *Flattening the earth: Two thousand years of map projections.* Chicago: University of Chicago Press.

- *Two Publications by the Same Author in the Same Year*

  List entries in alphabetical order by publication title. Then assign lowercase letters to each date of publication in order (2002a, 2002b, etc.).

  Long, J. R. (2000a). A 5.1–5.8 GHz low-power image-reject downconverter. *IEEE Journal of Solid-State Circuits, 35*, 1320–1328.

  Long, J. R. (2000b). Monolithic transformers for silicon RF IC design. *IEEE Journal of Solid-State Circuits, 35*, 1368–1382.

- *Book with Editor(s)*

  Dorfman, M., & Thayer, R. H. (Eds.). (1996). *Software engineering.* Toronto: John Wiley & Sons Canada.

- *Book Published in New Edition*

  Diamond, W. J. (2001). *Practical experiment designs for engineers and scientists* (3rd ed.). Toronto: John Wiley & Sons Canada.

- *Book Published in Volumes*

  Verschueren, K. (2001). *Handbook of environmental data on organic chemicals* (4th ed., Vols. 1–2). Weimar, TX: Culinary and Hospitality Industry Publications Services.

- *Article in an Encyclopedia*

Hinrichs, T. (1992). Geothermal power. In *McGraw-Hill encyclopedia of science and technology* (7th ed., Vol. 8, pp. 83–87). New York: McGraw-Hill.

- *Article in a Scholarly Journal*

Brett, N. C. (1991). Language laws and collective rights. *The Canadian Journal of Law and Jurisprudence, 4*(2), 347–360.

- *Article in a Monthly Periodical*

Ricciardelli, A., Pizzimenti, D., & Mattei, M. (2003). Passive and active mass damper control of the response of tall buildings to wind gustiness. *Engineering Structures, 25*(9), 1199–1209.

- *Article in a Newspaper*

Palmer, K. (2003, May 7). West Nile plan outlined: Province vows to increase spending. *Toronto Star*, p. B5.

- *Article without an Author*

List the article alphabetically by title.

Biotech crops get third-world boost. (2003, July 11). *The Globe and Mail*, p. C4.

- *Video Recording*

List alphabetically by director or producer. This form of entry can also be used for audiotape, slides, and film.

Coast Community College District (Producer). (1990). *Decision making and problem solving* [Videocassette]. Toronto: TVOntario.

- *Article in an Online Journal*

Wang, F., Juniper, S. K., Pelegrí, S. P., & Macko, S. A. (2003). Denitrification in sediments of the Laurentian Trough, St. Lawrence Estuary, Quebec, Canada. *Estuarine, Coastal and Shelf Science, 57*(3), 515–522. Retrieved July 16, 2003, from http://www.sciencedirect.com/science/journal/02727714

- *CD-ROM*

Nichols, M. (1995, March 13). High-tech artificial limbs. *Maclean's*. Retrieved October 23, 2000, from *The Canadian Encyclopedia 2001* [CD-ROM]. Toronto: McClelland & Stewart.

- *E-mail*

Do not include personal communications in an APA reference list.

- *Web Page*

Maddison, D. R., Maddison, W. P., Schulz, K. S., Wheeler, T., & Frumkin, J. (2001). *The Tree of Life Web Project* [Online]. Retrieved July 15, 2003, from http://tolweb.org

- *Article from a Subscription Service or Online Database*

Bergman, B. (2003, June 23). Born to be high and wild. *Maclean's* 116(25). Retrieved September 30, 2003, from *EBSCOhost Academic Search Premier*. Rogers Media Publishing Ltd. Item 10046413.

MODERN LANGUAGE ASSOCIATION (MLA)

The Modern Language Association has been publishing style guidelines for scholars in modern languages since 1951. Over the years, the MLA method of documentation has changed from footnotes and bibliography to parenthetical citations and works cited. The 6th Edition of the *MLA Handbook for Writers of Research Papers* was published in 2003.

MLA Parenthetical Citations

An MLA parenthetical citation consists of the first word used to identify the source in the Works Cited list, usually the author's surname, and the page number where the information for the report was found in the source, if one is available. It looks like this: (Smith 328). Note that each citation is enclosed in round brackets and contains no punctuation.

Forms for MLA Parenthetical Citations

- *One Author.* Author and page: (Smith 328).

- *Two Authors.* Put them both in the citation: (Smith and Jones 328).
- *Three or More Authors.* Write all authors' names, or use the first name only: (Smith, Jones, and Tremblay 328) *or* (Smith 328).
- *Book Published in Several Volumes.* Include the volume number in the citation: (Smith 2: 328).
- *Corporate Author.* (Texas Instruments 16)
- *No Author.* Cite by first word of title: ("Joint" 194). Italicize the titles of books; place the titles of articles in quotation marks. (See Box 5.7.)

Forms for MLA Works Cited Lists

List all the sources you cited in the text of your report alphabetically by author surname under the centred heading Works Cited (see Box 5.8). Set up the entries with hanging indentation and double spacing. Sometimes you will read published sources that prepare you to write your paper but that you do not incorporate into your report in any way. These are uncited sources. Some report readers may wish to dig deeper into the subject of your report beyond the list of cited sources. Therefore, you may add a list of uncited sources after the Works Cited list, under the title Works Consulted. Use the same format as the Works Cited list.

If you have more than one source by the same author, list the sources alphabetically by title. For the second entry, use three hyphens in place of the author's name:

Smith, Kevin. *Book A.* . . .

---. *Book B.* . . .

Place the titles of independent works, such as books and journals, in italic typeface or underline them. Capitalize each significant word in the title. Place the date of publication following the publisher's information.

For an authored chapter in an edited book, list the author and title of the chapter first. Link the chapter to the book with the word "In." Then list the editor, title, and other details of the book.

Enclose Universal Resource Locators (URLs) for Web pages with angle brackets: < and >.

10

The most important monument at Copán is Altar Q (Coe 174). This limestone slab carries glyphs on all four sides and on the top. It provides us with some details about the earliest events in the reign of the first king, Kinich Yax Kuk Mo, as well as the dynastic succession of all the rulers of Copán. In one inscription, Kinich Yax Kuk Mo is said to celebrate the great, period-ending date in the Mayan calendar, 9.0.0.0.0, or December 11, 435 A.D. In all, Altar Q carries information concerning sixteen rulers of the city. As epigrapher David Stuart (16) has concluded, "The last date at Copán is associated with the last ruler, Yax Pasah. This is July 24th, 805 A.D., after which the record is silent."

Box 5.7: MLA Parenthetical Citations

14

WORKS CITED

Coe, Michael D. “From Huaquero to Connoisseur: The Early Market in Pre-Columbian Art.” *Collecting the Pre-Columbian Past.* Ed. E.H. Boone. Washington, DC: Dunbarton Oaks, 1993. 271–90.

GB Online. “Maya codices.” *GB Online’s Mesoamerica.* 2002. 20 Mar. 2004 <http: //pages.prodigy.com/GBonline/mesowelc.html>.

Hammond, Norman. *The Maya.* New Jersey: Rutgers UP, 1982.

Box 5.8: MLA Works Cited List

Examples of MLA Works Cited Reference Entries

- *Single Author*

Erjavec, Jack. *Automotive Technology: A Systems Approach.* 3rd ed. New York: Delmar Thomson Learning, 2000.

- *Two Authors*

McNair, Harold M., and James M. Miller. *Basic Gas Chromatography: Techniques in Analytical Chemistry.* New York: Wiley-Interscience, 1997.

- *Multiple Authors*

Griffiths, Anthony J. F., Jeffery H. Miller, David T. Suzuki, Richard C. Lewontin, and William M. Gelbart. *An Introduction to Genetic Analysis.* 6th ed. New York: W. H. Freeman, 1996.

- *Two Publications by the Same Author*

List entries in alphabetical order by title. Insert three hyphens in place of the author's name in the second listing to show that it is the same author as the work above it.

Snyder, John P. *Flattening the Earth: Two Thousand Years of Map Projections.* Chicago: U of Chicago P, 1993.

---. *Map Projections: A Working Manual.* U.S. Geological Survey Professional Paper 1395. Washington, DC: USGPO, 1987.

- *Book with Editor(s)*

Dorfman, Merlin, and Richard H. Thayer, eds. *Software Engineering.* Toronto: John Wiley & Sons Canada, 1996.

- *Book Published in New Edition*

Diamond, William J. *Practical Experiment Designs for Engineers and Scientists.* 3rd ed. Toronto: John Wiley & Sons Canada, 2001.

- *Book Published in Volumes*

Verschueren, Karen. *Handbook of Environmental Data on Organic Chemicals.* 4th ed. 2 vols. Weimar, TX: Culinary and Hospitality Industry Publications Services, 2001.

- *Article in an Encyclopedia*

Hinrichs, Thomas. "Geothermal Power." *McGraw-Hill Encyclopedia of Science and Technology.* 7th ed. (Vol. 8, pp. 83–87). New York: McGraw-Hill, 1992.

- *Article in a Scholarly Journal*

Brett, Nathan C. "Language Laws and Collective Rights." *The Canadian Journal of Law and Jurisprudence* 4.2 (1991): 347–360.

- *Article in a Monthly Periodical*

Ricciardelli, A., D. Pizzimenti, and M. Mattei. "Passive and Active Mass Damper Control of the Response of Tall Buildings to Wind Gustiness." *Engineering Structures* 25.9 (2003): 1199–1209.

- *Article in a Newspaper*

Palmer, Karen. "West Nile Plan Outlined: Province Vows to Increase Spending." *Toronto Star* 7 May 2003: B5.

- *Article without an Author*

List the article alphabetically by title.

"Biotech Crops Get Third-World Boost." *Globe and Mail* 11 July 2003: C4.

- *Video Recording*

List alphabetically by title.

Decision Making and Problem Solving. Coast Community College District. Videocassette. Toronto: TVOntario, 1990.

- *Article in an Online Journal*

Wang, Fenghai, S. Kim Juniper, Silvia P. Pelegrí, and Stephen A. Macko. "Denitrification in Sediments of the Laurentian Trough, St. Lawrence Estuary, Quebec, Canada." *Estuarine, Coastal and Shelf Science* 57.3 (2003): 515–522. Science Direct. 16 July 2003 <http://www.sciencedirect.com/science/journal/02727714>.

- *CD-ROM*

Nichols, Mark. “High-tech Artificial Limbs.” *Maclean’s.* 13 Mar. 1995. In *The Canadian Encyclopedia 2001*. CD-ROM. Toronto: McClelland & Stewart, 2000.

- *E-mail*

Birnstihl, Jennifer. “Re: Update.” E-mail to Professor C. L. Gulston. 12 May 2003.

- *Web Page*

Maddison, David R., Wayne P. Maddison, Katja S. Schulz, Travis Wheeler, and Jeremy Frumkin. *The Tree of Life Web Project.* 15 July 2003 <http://tolweb.org>.

- *Article from a Subscription Service or Online Database*

Bergman, Brian. “Born to Be High and Wild.” Maclean’s 116 (25) 23 June 2003. *EBSCOhost Academic Search Premier.* Rogers Media Publishing Ltd. Item 10046413. Sir Sandford Fleming College LRC. 30 Sept. 2003 <http://fleming0.flemingc.on.ca/lrc/library/libav2.htm>.

CHECKLIST FOR REPORT DOCUMENTATION

Be sure that you have . . .

- Chosen the correct system: CSE, APA, MLA, or other
- Cited and referenced sources used in text

CSE

Name-Year System

- Placed a citation after each borrowed text passage and graphic presentation in the report
- Placed citations in parentheses, with author surname and year of publication: (Smith, 2004)
- Called the reference list References or References Cited
- Organized the References Cited list in alphabetical order by author surname
- Used hanging indentation format, single spaced
- Ensured all entries have complete information and correct order of items

Citation-Sequence System

- Placed citation after each borrowed text passage and graphic presentation in the report
- Used Arabic numerals for citations, in parentheses: (1), (2), etc.
- Called the reference list References or References Cited
- Arranged the reference list in numerical order, single spaced
- Ensured all entries have complete information and correct order of items

APA

- Placed a citation after each borrowed text passage and graphic presentation in the report
- Put citations in parentheses, with author surname and publication year: (Smith, 2004)
- Included page number for direct quotations cited: (Smith, 2004, p. 105)
- Called the reference list References
- Listed sources in alphabetical order
- Double-spaced reference entries, using block or indented paragraph format
- Ensured all entries have complete information and correct order of items

MLA

- Placed a citation after each borrowed text passage and graphic presentation in the report
- Put citations in parentheses including author's name or first word from Works Cited list and page number
- Called the reference list Works Cited
- Checked to see if additional Works Consulted list is required
- Listed sources in alphabetical order
- Double-spaced reference entries
- Ensured whole publication titles are in italics or underlined; partial publication titles in quotation marks
- Ensured all entries have complete information and correct order of items

APPENDIX A: BUSINESS ELEMENTS OF A REPORT

Memo or Letter of Transmittal

The purpose of this element is to transmit your report formally to the person or group of people named as the destination on the title page. Its content summarizes the purpose and content of the report, and it transfers responsibility for the report's contents to the reader. Place this element after the title page and before the table of contents. Choose the memo form if the report audience is within your company or agency; choose the letter form if the report audience is outside your company or agency.

Write three paragraphs summarizing the purpose and scope of the study, its methodology and major findings, and your conclusions and recommendations. The last paragraph may also cover follow-up to the report and provide contact information, depending on the workplace situation (see Box A.1).

MEMORANDUM

To: Prof. C. L. Gulston
From: Jonathan White
Subject: Short-rotation hybrid poplar farming
Date: April 19, 2004

I am pleased to present you with my report on the growth and utilization of poplar in Canada. I am submitting this report in partial fulfillment of the requirements of the Silviculture course in the Forestry Program at Sir Sandford Fleming College.

The following report describes general growth processes in woody stems and then examines these growth processes in poplar. The conclusion emphasizes the opportunities in Canada to take advantage of poplar's wide range, high volume, and fast growth by developing short-rotation poplar farms for veneer, plywood, and pulpwood.

The silvicultural plan set out in this report should stimulate discussion in our course as it has in the poplar industry. I will eagerly await your evaluation of this project. If you have questions, or need further information, please e-mail me at jwhite@flemingc.on.ca or call me at (705) 555–1472.

Box A.1: Memo of Transmittal

Executive Summary

An executive summary is longer than an abstract, usually about one page for an average-length report (see Box A.2). It covers the methodology, results, and especially conclusions and recommendations in greater detail. Place the executive summary after the other elements of front matter and before the introduction. Some business reports place the executive summary directly in the introduction or at the end of the report after the conclusion.

EXECUTIVE SUMMARY

This report presents four main growth processes in woody stems:

- primary growth
- secondary growth
- formation of annual rings
- formation of heartwood and sapwood.

These primary processes determine the utilization of our forest products.

In poplar, primary growth occurs rapidly in root suckers and stump sprouts. Suckers grow faster than seedlings because of their developed root system. Secondary growth produces volume. Canadian poplar sites are similar to European sites that have produced good poplar volume in 4 to 6 years. This short rotation is good for poplar because of the prevalence of decay in mature trees. A rotation age of 30 to 50 years will offset problems of cull. The industry must scale operations for smaller-diameter logs.

The goal of experiments with mini-rotations of poplar is to produce hybrid poplar species with the following characteristics:

- frost-hardiness for northern sites
- high survival rate and growth of root suckers
- good height and diameter growth in root suckers and stump sprouts
- resistance to insects and diseases.

Box A.2: Executive Summary

APPENDIX B: MICROSOFT WORD XP TUTORIALS

These tutorials provide brief descriptions of techniques that you can use in Microsoft Word XP™ to achieve the correct format for specific elements found in most scientific and technical reports. They will not replace a thorough knowledge of the software. The Office Assistant for Word XP can also help you with descriptions of menus, commands, and functions. Most of the commands shown are similar to those required by Microsoft Word 97™ and Word 2000™.

Outlining

The Outline view in Word XP is a view of your entire document that allows you to move headings into different sections or different levels of organization in the hierarchy. Outline view has powerful features: with it, you can generate a table of contents, format report headings in a variety of styles, restructure an existing document, and so on. Do not confuse these implications of Outline view with the creation of an outline of report headings to organize your research notes before you begin to write.

To create a simple outline, refer to Figure B.1 and follow these steps:

1. From the Format menu, select the Bullets and Numbering option. You now have the Bullets and Numbering menu in a separate window.

2. Select the Outline Numbered tab on the top of the Bullets and Numbering menu. You may now select the style of numbering you wish for your outline. Complete the command by clicking OK at the bottom of the window.

3. If you wish to customize the numbering system for your outline, select the Customize button at the bottom of the Bullets and Numbering menu. You now have the Customize Outline Numbered List menu from which you may select your own outline number format, style, and position for each level of organization (see Figure B.2).

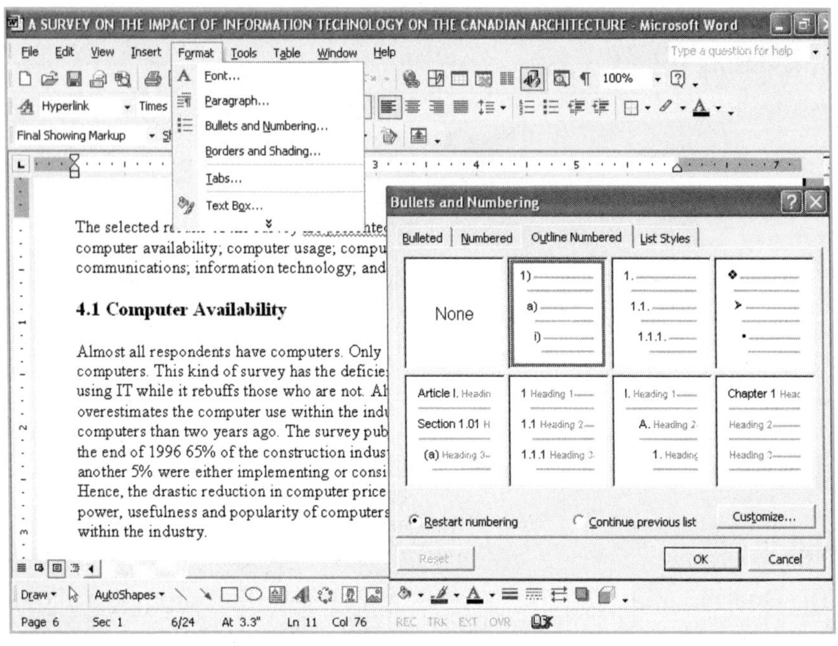

Figure B.1: Format Menu with Bullets and Numbering

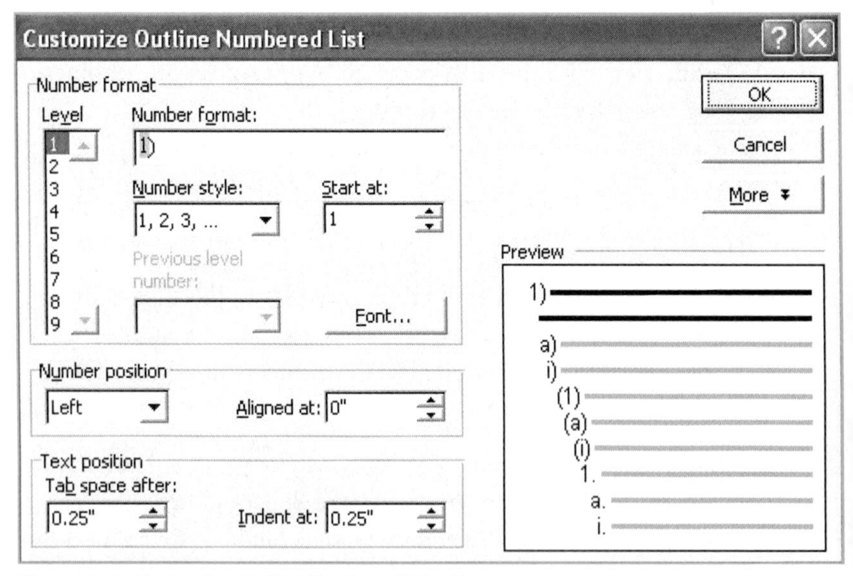

Figure B.2: Customize Outline Numbered List Menu

1. Use the Tab key to move your cursor to the next lower level of your outline. Use Shift + Tab to move your cursor to the next higher level of your outline.

2. An alternate method of formatting your outline is to select the List Styles tab on the Bullets and Numbering menu. Like the Outline Numbered menu, you have the option of customizing the number and heading styles of your outline.

Inserting Page Numbers

Pagination is complex in Word. To avoid the difficulties of inserting section and page breaks in a single file with all report elements, create two files: one for the front matter and another for the report body. This will allow you to create two page-number sequences easily. If you have an extensive appendix, create a third file for that. Figure B.3 shows Word's page-numbering menus.

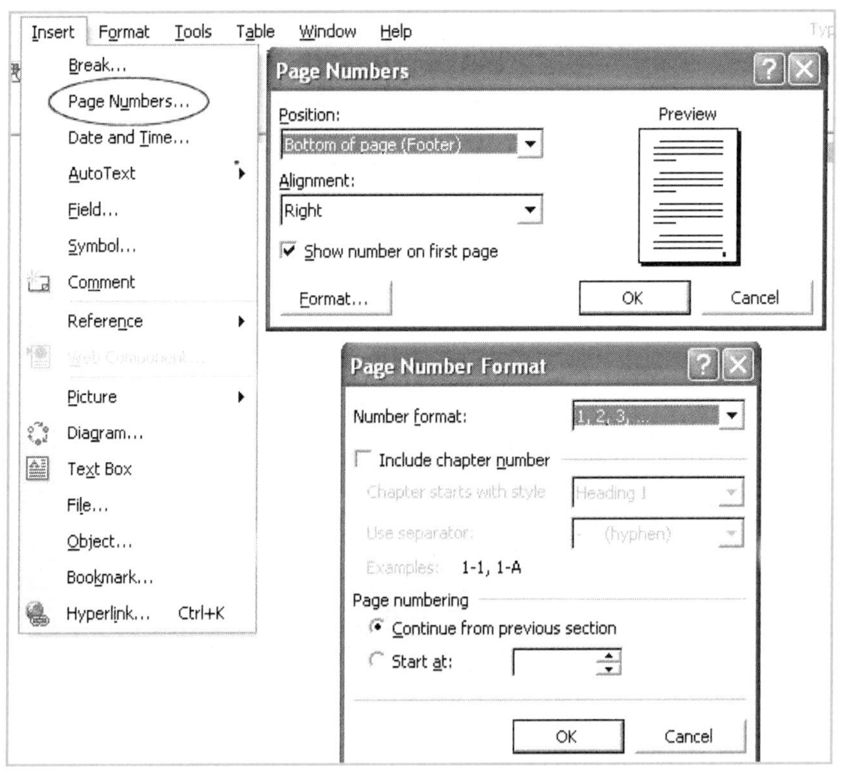

Figure B.3: Page Numbering Menus

Page Numbers for Front Matter

1. In your Word file, include all the elements of the front matter that your report requires, from the title page to the abstract, in their correct order. See the section on formal report elements.
2. To insert page numbers, click on the Insert menu above the standard toolbar at the top of the page.
3. Select the Page Numbers command from the Insert menu and click on it. The Page Numbers window will pop up.
4. In the Page Numbers window, select Position: Bottom of Page (Footer) and Alignment: Center. This will centre your page numbers for the front matter at the bottom of each page.
5. Uncheck the box in the lower right marked "Show number on first page." This will remove the page number from the title page.
6. Click on the Format button. This gives you the Page Number Format menu. In the Number Format window, select lower-case Roman numerals.
7. To complete the command, click OK in the Page Number Format and Page Numbers windows.

Page Numbers for Report Body

1. Include all the pages of the body that your report requires, from the Introduction to the References page, in the correct order in one file.
2. To insert page numbers, click on the Insert menu above the standard toolbar. Select the Page Numbers command. The Page Numbers menu will pop up. From the Page Numbers menu, select Position: Top of Page (Header) and Alignment: Right. This will place your page numbers for the body in the top right corner of each page.
3. Leave checked the box in the lower right marked "Show number on first page." This will start the page numbering from the Introduction page. If your report format requires no page number on the first page, uncheck the box.
4. To complete the command, click OK in the Page Numbers window.

Note: The default Arabic page numbers are in 10-point Times New Roman typeface. If you wish to change to these to match your report base font, double-click the page number on the first page, highlight it with your mouse, and use the Font command on the Format menu to change the font.

Inserting Leader Dots in the Table of Contents

1. Place your mouse cursor on the first line with a heading and a page number, and click to get a flashing text cursor. This is necessary because Word formats the line with the flashing text cursor. Be sure your Ruler line is showing. This can be selected from the View menu.
2. To insert leader dots from the heading to the page number, you must first set the correct tab stop where you wish to place your page number column. Click on Format above the standard toolbar at the top of the page.
3. Select the Tabs command.
4. In the Tabs window, type in the value for the tab-stop position where you want the page numbers. This value can be estimated from the Ruler line.
5. For this special tab, select Right from the Alignment menu. Remember that the page-number column must be on the right side of the Contents page.
6. To insert leader dots from the heading to the page number tab, select item 2 from the Leader menu. This style of leader dots is the standard used in technical reports.
7. Click OK to set the command.
8. To insert leader dots on the line, begin with your cursor at the end of the heading on the first line. Hit the Tab key. The cursor will move to the special tab position and insert leader dots. Type the page number. Repeat for each new line.
9. If you need to insert leader dots and page numbers in lines with headings already typed, use the Format Painter command. Activate Format Painter with the brush icon. With your cursor in a line with

the special tab and leader dots, double-click your mouse cursor on the brush icon. Doing this identifies the line with formatting that you want to replicate elsewhere in the document. Move your mouse cursor to the line where you want to have leader dots. Notice that the mouse cursor has changed to a brush icon. Left-click in each line where you wish to have the special tab set and the leader dots. When you are finished, click the mouse cursor on the brush icon again to turn it off. See Figure B.4.

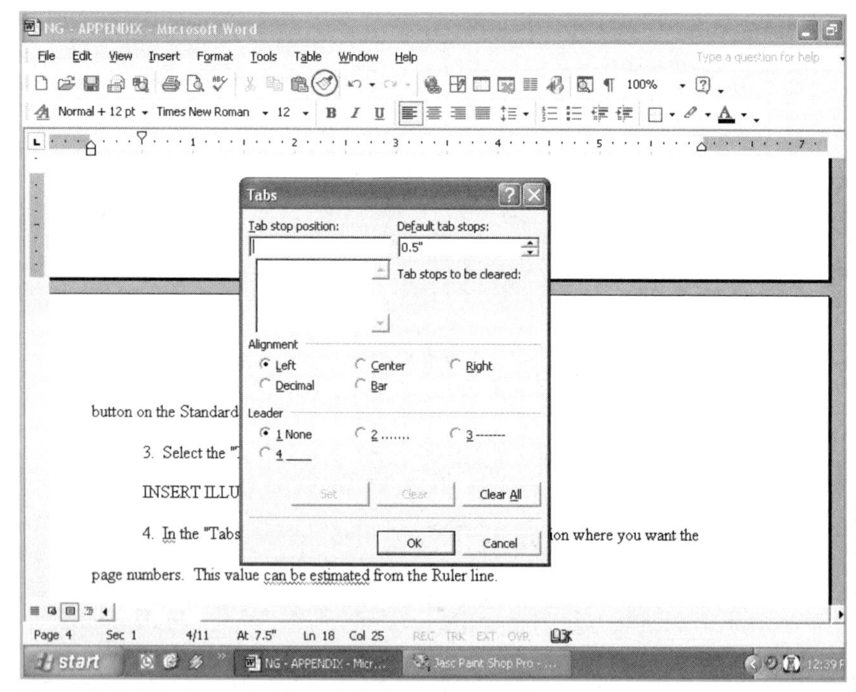

Figure B.4: Tabs Menu and Format Painter Icon

Wrapping Text around Figures

This tutorial assumes that you know how to acquire graphic images in Microsoft Word. You can bring graphics into a Word file using the Picture command on the Insert menu above the standard toolbar. You can also copy the image from other software, such as an Internet Web browser or Adobe Illustrator graphic software, to the Windows clipboard and paste it into Word.

Microsoft Word allows you to adjust a graphic image on your report page. These adjustments include image control such as greyscaling or watermarking, contrast and brightness controls, resizing and rotating, cropping, adding borders, compressing graphics for Web display, and page-layout formatting. You can find these adjustments on the Picture toolbar. You can also move the image around on your page by clicking and holding the left mouse button while you move the image. When you insert a graphic image on a text page, the image appears wrapped in line with text. This is the default setting and may be a suitable page layout for your image. If you want a different text-wrapping style for your picture layout, you must select an alternate wrapping mode. Figure B.5 illustrates the following steps on selecting and adjusting a text-wrapping style.

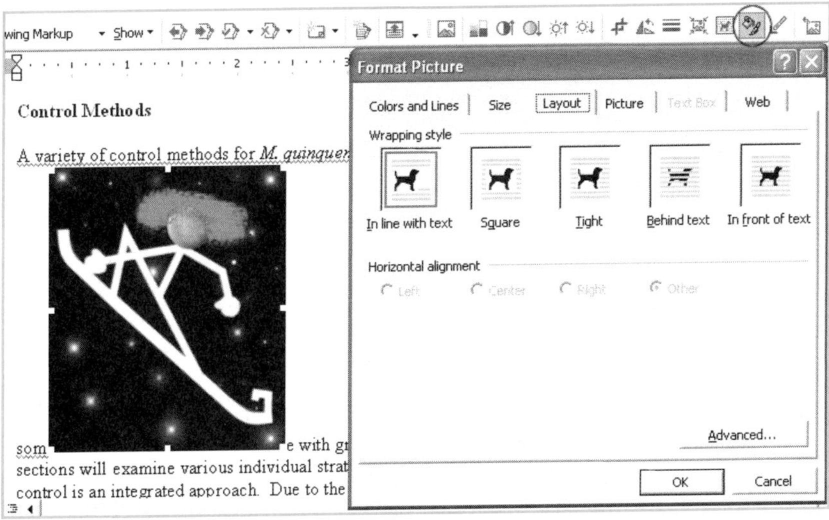

Figure B.5: Selected Picture with Handlebars, Circled Picture Format Icon, and Picture Format Menu

1. Select your image by left-clicking on it. Little square handlebars will appear along the edges of the image. These allow you to resize the image by clicking and dragging the handlebars with your mouse.

2. With the image selected, you will now see the Format Picture icons on the toolbar above. Note that you can also right-click the image to produce an instant menu from which you can select the Format Picture commands.

3. Click on the Format Picture icon on the Format Picture toolbar (the circled icon in Figure B.5). This will bring up the Format Picture menu in a new window.

4. In the Format Picture window, left-click on the Layout tab.

5. Choose from the wrapping styles shown by left-clicking on the appropriate icon. Four of the five options also allow you to select the horizontal alignment of your graphic image on the text page.

6. The Advanced button on the Format Picture menu will bring up the Advanced Layout options in a window (see Figure B.6). These controls will allow you to adjust the text wrapping and picture position of your image more exactly.

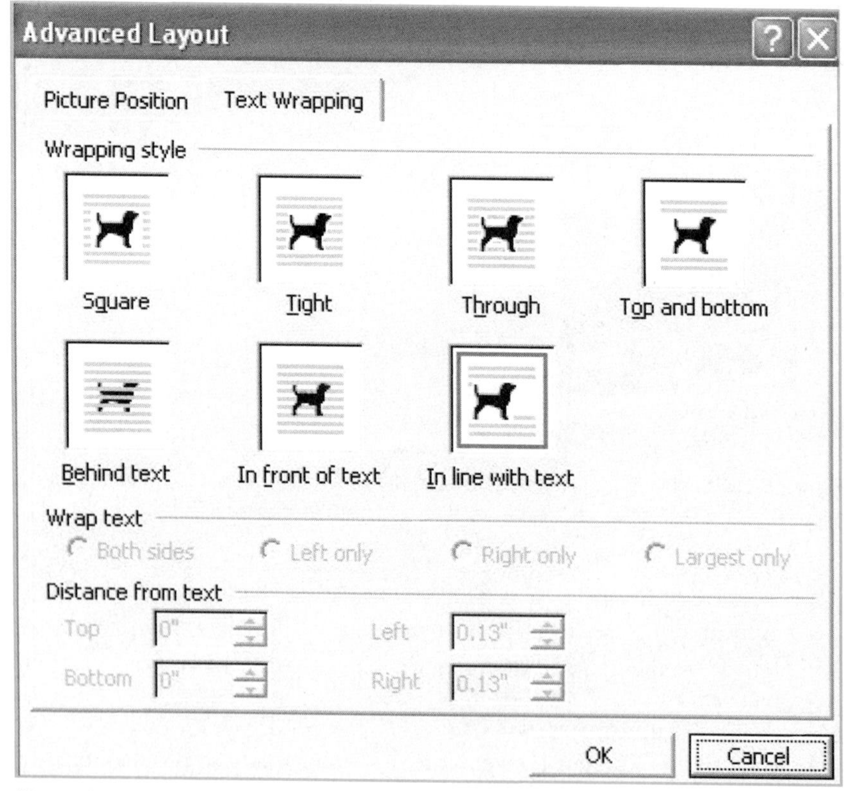

Figure B.6: Advanced Layout Menu

Note: For large images, we recommend the "Top and Bottom" layout. Give your images room on the page. For smaller images, use Square or Tight. The Advanced Layout controls will allow you to increase the distance of the image above and below your report text. Experiment with your image and text to get the right amount of white space around your image.

Inserting Figure Number and Title for Small Figures

For a large image with the Top and Bottom layout, place the required figure number and title below the figure using regular text. For a small image using the Square or Tight layout, insert the required figure number and title with the Text Box function as described below.

The text box is actually a graphic image that contains text. Treat it as you would an image, including using its handles to resize it and text wrap to arrange text around it. Within the box, choose the same font as your report text by selecting the text box and choosing the appropriate style and size of font in the window on the Formatting toolbar.

1. Select Text Box either from the Insert menu above the standard toolbar or from the Drawing toolbar at the bottom of your Microsoft Word XP screen. Figure B.7 shows the icon for the Text Box circled on the Drawing toolbar. Be sure that you have already activated the Drawing toolbar by selecting it from the Toolbars menu on the View menu above the standard toolbar.

2. When you have selected Text Box by clicking on the icon, your mouse cursor will change to a + shape. Word XP will also insert a drawing canvas, a grey box that says, "Create your drawing here." This box is for drawing multiple shapes that you can move or resize as a group. The drawing canvas is unnecessary in this application; remove it by pressing the Delete key on your keyboard.

3. To create a text box for your figure number and title, place the + mouse cursor immediately below your small image, aligned with its left side. Click and drag the text box into a rectangle the width of your image. (See Figure B.8.)

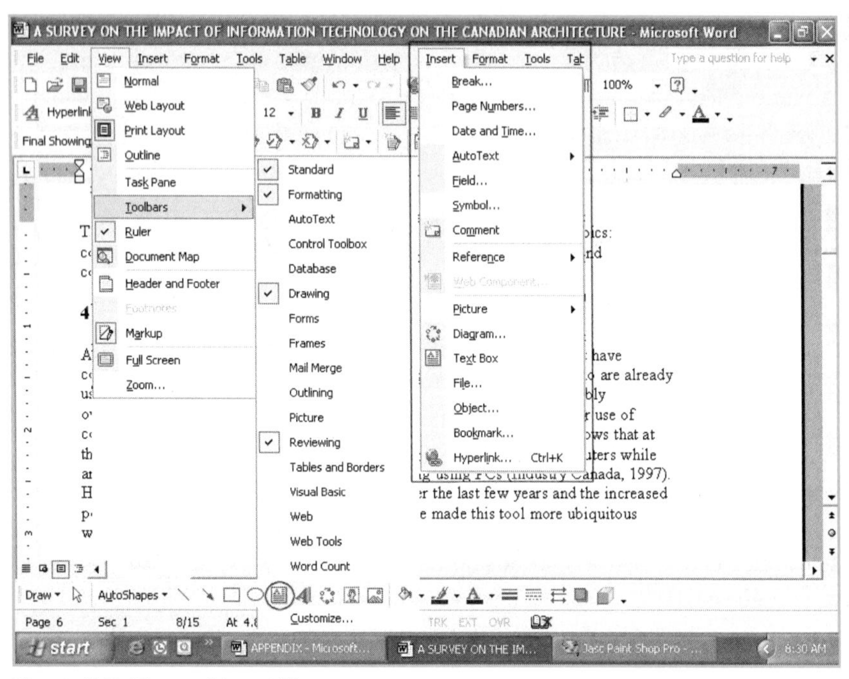

Figure B.7: View and Insert Menus

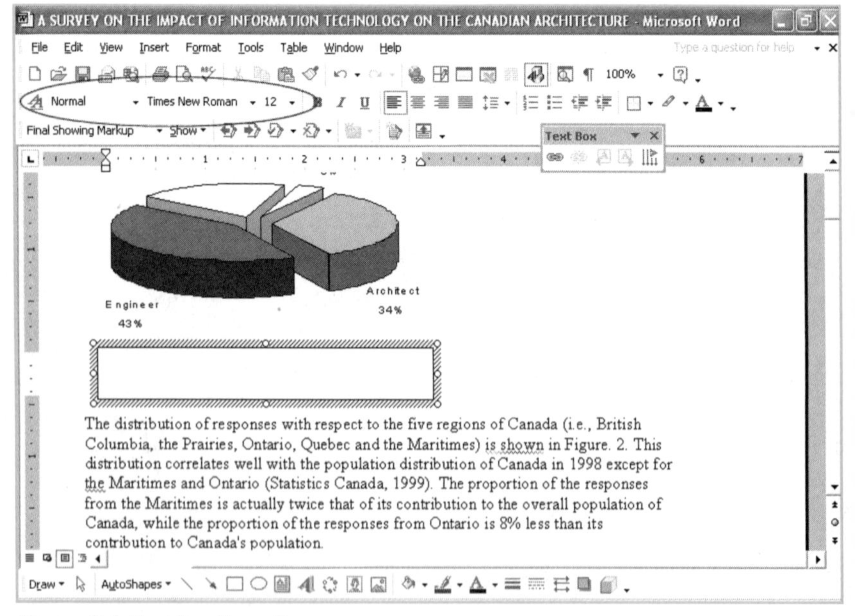

Figure B.8: Text Box Formation

4. Use the handlebars on the new text box to resize it into a rectangle that is the width of your image and deep enough to accommodate the text of your figure number and title. From the font window on the Formatting Toolbar (circled in Figure B.8), select the style and size of your report base font. Type an appropriate figure number and title for your image. Add an appropriate citation if you borrowed the graphic from a secondary source.

5. Refer to Figure B.9 for the following steps to make your text wrap around the text box. Select your text box by left-clicking on it; the outline of the rectangle should be surrounded with *hachures* (light, angled lines) and contain handlebars for resizing the box.

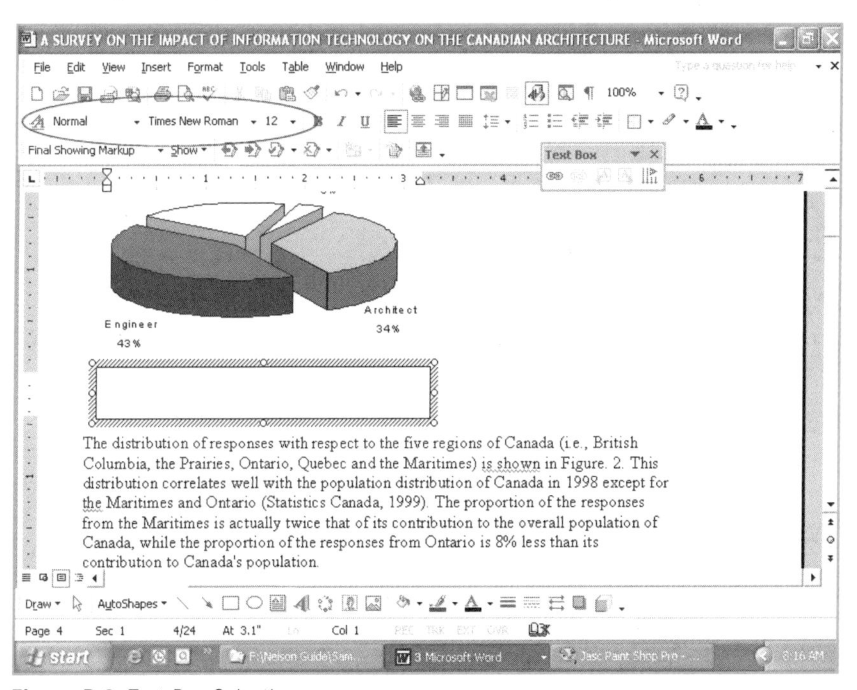

Figure B.9: Text Box Selection

6. Either select the Text Box command from the Format menu or place your mouse cursor on the Text Box rectangle and right-click. This brings up a menu that includes the Format Text Box command. Either selection will give you the Format Text Box menu in a new window.

7. In the Format Text Box window, click the Layout tab. Choose the same method of wrapping your report text around the text box as you used for the figure itself. The default text box has a solid black line outlining the text. This is not appropriate in report writing. To eliminate the solid line, click the Colors and Lines tab (see Figure B.9).

8. In the Line area of the Colors and Lines menu, click the down arrow on the right side of the Color rectangle. The colour palette will appear with coloured squares.

9. Choose white in the lower right corner. This changes the colour of the text box outline from black to white, thus making it invisible in your report. Click OK to execute the command.

Note: You can still click on the text box to select it after it is outlined in white. This allows you to continue to resize the box or to reposition the text box below your image.

Applying Hanging Indentation for Reference Lists

Format reference lists using hanging indentation. This makes surnames easier for your reader to see. The first line of each entry begins on the left margin. The second and successive lines in hanging indentation begin about five spaces from the left margin.

Microsoft Word XP considers each entry in your list to be a paragraph. The default paragraph form begins on the default left margin of your page. To change the paragraph form to hanging indentation for your reference list entries, follow these steps.

1. Click your mouse cursor anywhere in the first entry of your reference list. This places the flashing cursor in that entry. The paragraph format command applies only to the paragraph where your flashing cursor is located when you activate the command.

2. To activate the Paragraph menu, select Paragraph from the Format menu (see Figure B.7). This will bring up the Paragraph menu in a separate window.

3. Choose the Indents and Spacing tab (see Figure B.10). In the Indentation area of the window, click the down arrow on the right side of the Special window.

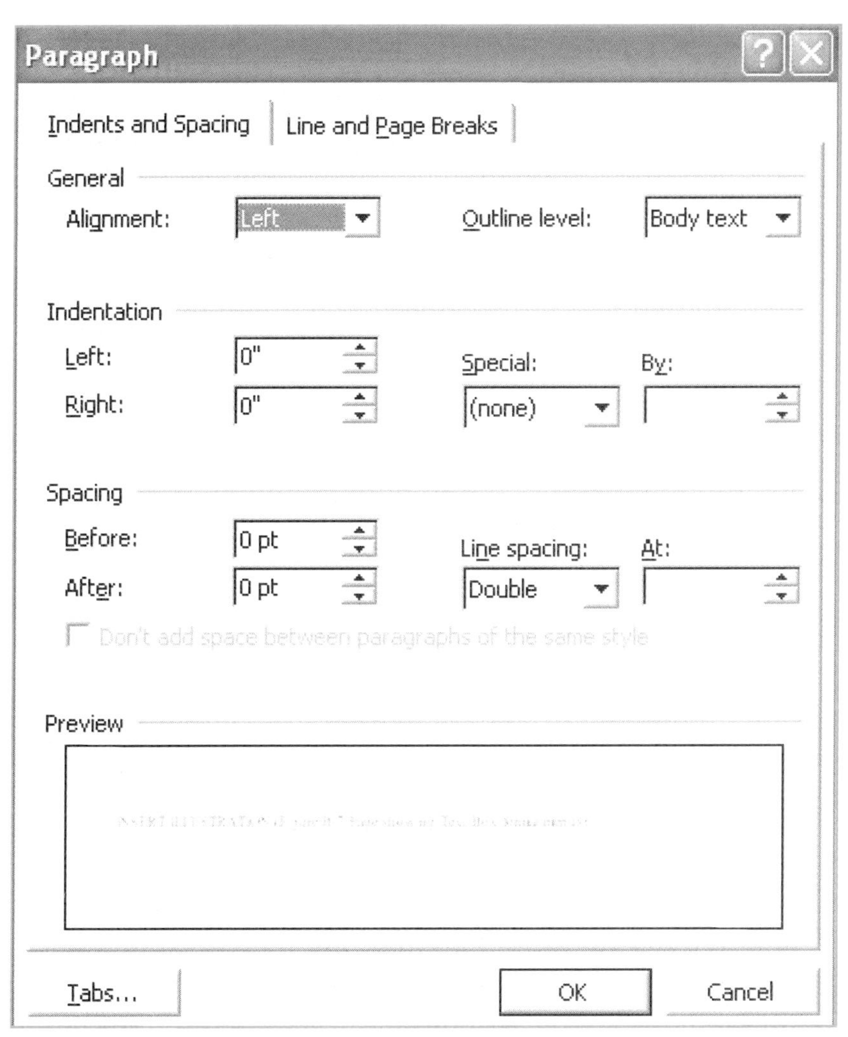

Figure B.10: Paragraph Menu

4. From the Special menu, select Hanging. If you have your line spacing set to Double as required in the body of your report, change this now to Single if required by your reference list. Click OK to execute the command.

5. Use the Format Painter (see Figure B.4) to transfer the hanging indentation format from the paragraph you just formatted to the other entries in your list.

Inserting Mathematical Equations

Insert mathematical equations into your Word document using Microsoft Equation Editor 3.0. You must first install this feature of Microsoft Office XP on your computer. For the following steps, refer to Figure B.11.

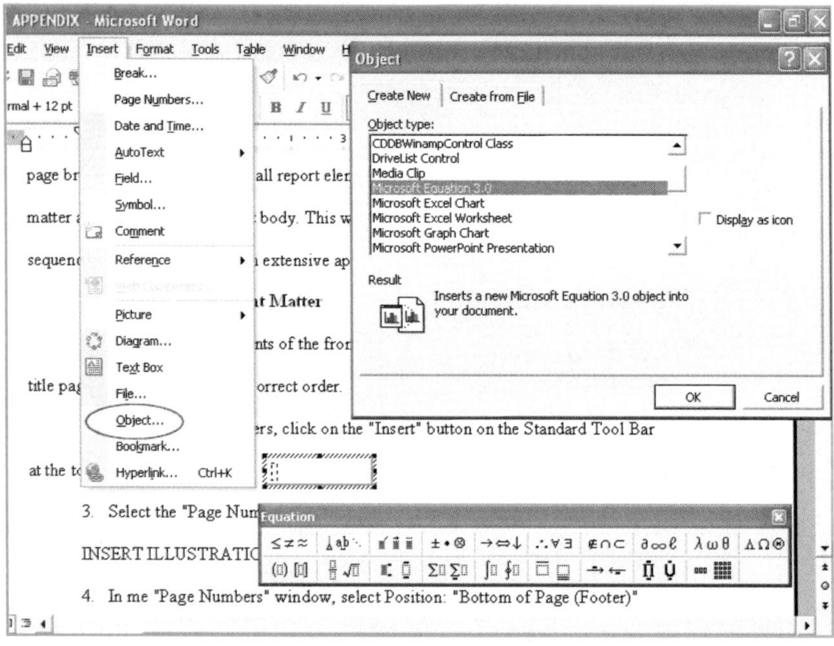

Figure B.11: Object and Equation Menus

1. Click your mouse cursor to place the flashing cursor in your text at the location where you want to insert a mathematical equation. Word XP will create an object box like a text box at that location, into which you can insert mathematical symbols to make equations.

2. From the Insert menu, select Object. This gives you the Object window, from which you can select Microsoft Equation 3.0 by highlighting it and clicking the OK button.

3. You now have the equation object box in your text and the Equation menu as a window floating above your text. Notice that the flashing cursor now appears in the equation object box within a rectangle defined by a dotted line. The symbols and numerical values that you insert will appear within the rectangle.

4. Insert mathematical symbols by selecting them from the Equation menu. Insert numbers and other characters directly from the keyboard. Select, move, resize, and text-wrap the equation object box just like the text box to suit your page requirements.

APPENDIX C: SAMPLE STUDENT REPORT

Note: The following sample report is intended only as a model of correct report format as set out in this text, using the CSE name-year style of documentation.

THE STRUCTURE AND GROWTH OF TREE STEMS AND THEIR IMPORTANCE TO THE GROWTH AND UTILIZATION OF POPLAR (*POPULUS*) IN CANADA

By
Jonathan White

To
Professor C. L. Gulston

April 10, 2004

TABLE OF CONTENTS

LIST OF ILLUSTRATIONS

iii

ABSTRACT

Primary growth and secondary growth of stems are the two most important processes that contribute to the total growth of trees. They are the major determining factors of utilization of our forest products. Primary growth is responsible for tree height, while secondary growth produces greater diameter.

This paper outlines briefly the main structures of woody stems and describes the processes determining their growth. It relates these processes to the utilization of poplar (*Populus*) in Canada. It identifies conditions that limit mature stem growth and summarizes research into improvement of stem growth and management of environmental factors affecting poplar silviculture.

1

INTRODUCTION

Purpose and Scope

The purpose of this report is to describe the growth processes of woody stems in simple terms to provide a framework for discussing the growth characteristics of poplar (*Populus*) in Canada with a view to applying these to poplar management. These data are a summary of well-established processes of tree physiology and current research into short-rotation poplar management.

The following are the four main growth processes presented in this report: primary growth, secondary growth, annual rings, and the formation of heartwood and sapwood. This report then examines these processes in poplar (*Populus*) species in Canada. This includes large and small-toothed aspen, cottonwood, and European hybrid species.

The poplar section of the report describes primary and secondary growth processes in poplar species and factors affecting each type of growth. It ends with a discussion of short-rotation poplar stands to take advantage of fast growth and minimize the effect of pests and disease in mature stands.

2

Review of Literature

Any secondary school or post-secondary school textbook of botany, plant physiology, or plant growth is suitable for gaining insight into the primary and secondary growth of stems. However, a particularly useful one quoted throughout this paper is Botany of Woody Stems by Luger and Jarrell, 2001.

A good summary of the growth and utilization of poplar is the proceedings of a poplar symposium that took place in Cornwall, Ontario, and Syracuse, New York, in 1987 at a joint meeting of the Poplar Councils of Canada and the United States. The publication, edited by Fayle et al. (1979) is entitled Poplar Research, Management and Utilization in Canada.

For a complete review of the literature, see Palmer's Short-rotation Culture of Populus and Larix: A Literature Review (1991).

3

TREE GROWTH PROCESSES

Primary Growth in Stems

The stem is the structure, usually above the ground, which bears the leaves and branches of the tree (Shigo, 1994: 53). Stems vary greatly in structure; each species differs in the way it combines cells into complex systems based on the vascular tissues that bind the stem together, conduct moisture, and give it strength.

Structures

Growth in trees occurs only in specific tissues called meristems or meristematic tissue. In the meristematic tissue, under favourable conditions, new cells are continually being formed as some or all of the cells repeatedly divide (Browse, 1988: 401). The most important meristems in the tree are the apical-stem meristem and the vascular cambium, respectively related to primary and secondary growth (Browse, 1988: 401).

Growth Sequence

There are three regions or zones in the apical stem, or tip of the terminal bud of the tree, which all play their parts during primary growth, or growth in stem length.

The region at the very tip of the stem, the first 1/4 inch, is the meristematic area (see Figure 1). The leaf primordium protects this area of active cell division or mitosis. Below this area is the zone of cell elongation, the only area where upward growth takes place (Luger and Jarrell, 2001: 126). In this zone, the undifferentiated parenchyma cells absorb water and cause the cells to enlarge vertically. The lowest is the zone of cell maturation where the now specialized or differentiated parenchyma cells begin to form the primary tissues: xylem, phloem, and cambium. The entire process could be visualized as a stalagmite growing on the floor of a cave. Throughout the entire process of primary growth, the apical-stem meristematic tissue is continually producing new cells (Luger and Jarrell, 2001: 128).

5

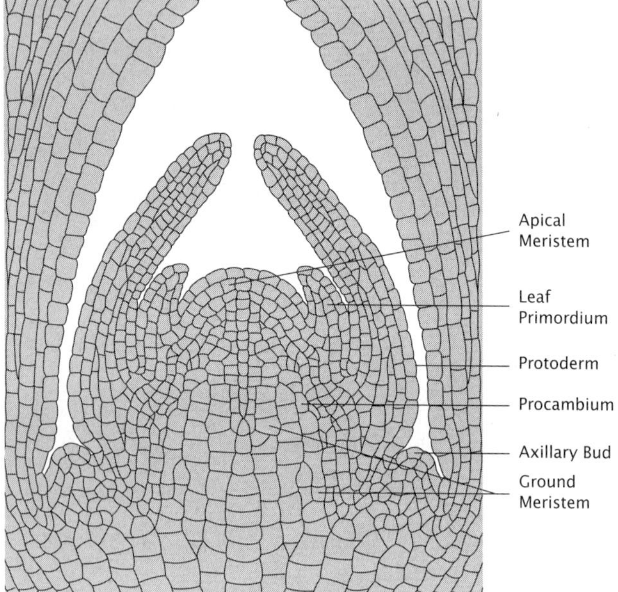

Figure 1. Regions of apical meristem (Luger and Jarrell, 2001: 253)

Secondary Growth in Stems

This results from the activity of the vascular cambium, which is also a form of meristematic tissue, and is characterized by an increase in stem thickness or diameter.

6

Structures

The cork cambium is a one-cell-thick layer between the bark and wood that repeatedly subdivides to form new wood and bark cells, or xylem and phloem (see Figure 2). The cork cambium usually becomes active before the primary vascular tissues have become fully differentiated (Luger and Jarrell, 2001: 132).

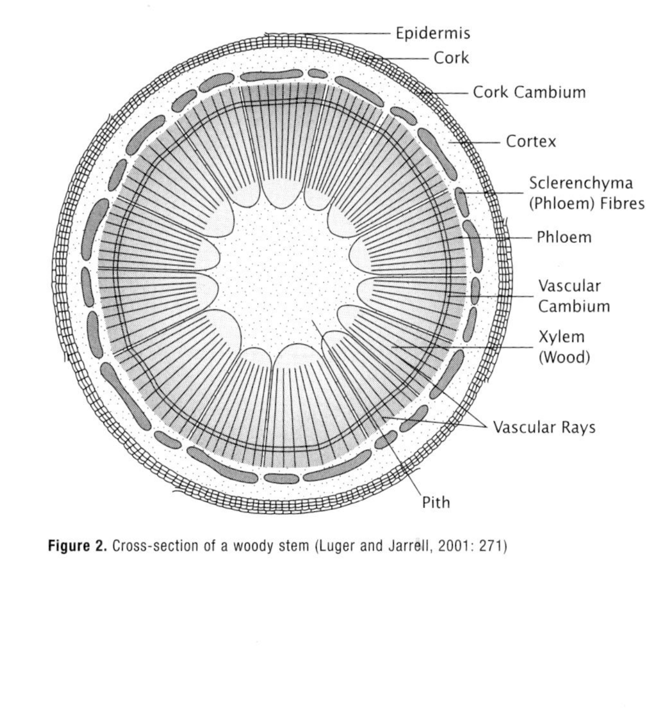

Figure 2. Cross-section of a woody stem (Luger and Jarrell, 2001: 271)

7

Growth Sequence

The cambium cells divide and produce primary xylem on the inner side and primary phloem on the outer side (see Figure 3) within the vascular bundles. They then lay down secondary xylem on the inner side of the primary xylem and secondary phloem on the inner side of the primary phloem (Luger and Jarrell, 2001: 139).

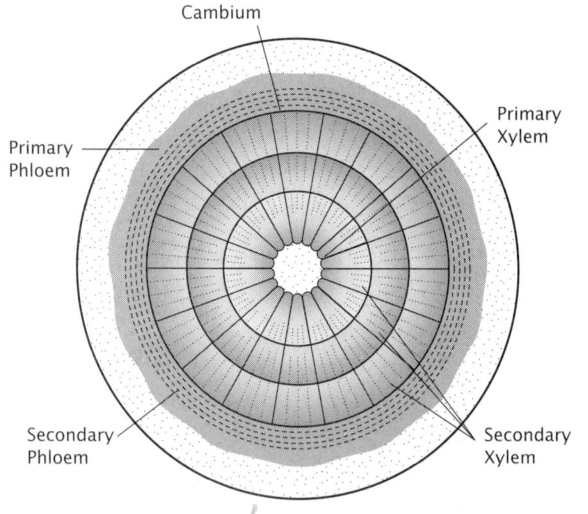

Figure 3. Primary and secondary growth cells (Luger and Jarrell, 2001: 152)

8

Annual Rings

By the end of the first growing season, the stem has its first annual growth ring (see Figure 4). This consists of a small amount of primary phloem next to the pith, the remainder of the ring being secondary xylem.

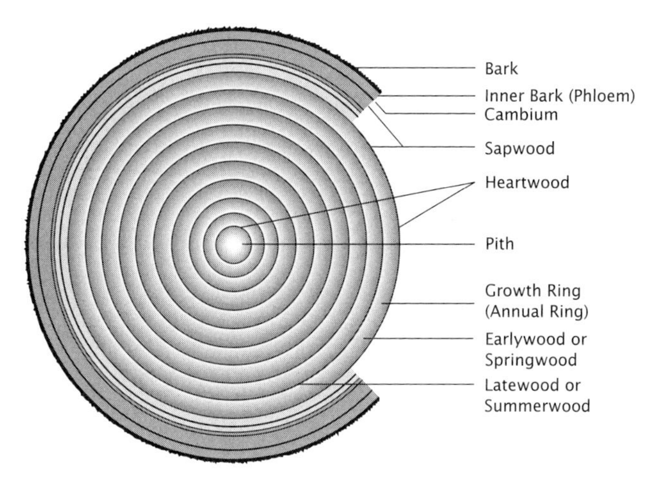

Figure 4. Formation of annual growth rings (Shigo, 1994: 24)

Primary and Secondary Xylem and Phloem

The cambium forms secondary xylem and secondary phloem year after year during the life of the tree (Luger and Jarrell, 2001: 164). The pressure of new tissue formed within the ring gradually crushes the older phloem cells but not the secondary xylem cells that eventually form the bulk of the stem. The compression of phloem cells also happens because of the faster rate of production of xylem cells, 75 to 100 times as fast (Luger and Jarrell, 2001: 165).

Springwood and Summerwood

Each annual ring contains springwood and summerwood (see Figure 4). Springwood is that portion of the annual growth ring formed during the early part of the growing season. In the spring when growth is fast, the primary growth cells are larger and thinner-walled than those produced in the summer. These cells carry water for developing leaves, shoots, and flowers (Luger and Jarrell, 2001: 165).

The number and size of summerwood cells exceed those of springwood because the summer growing season is longer than that of spring. The width of each annual ring reveals the environmental influences on a tree's growth for that year. Cooler, more moist conditions produce wider rings.

10

Formation of Heartwood and Sapwood

Conversion of sapwood into heartwood as xylem tissue increases in age is the result of changes in colour, composition, and structure of various elements. Sapwood is the living wood of pale colour near the outside of the stem and is susceptible to decay. Sapwood cells carry oxygen and mineral sap up the tree (Browse, 1988: 132). The inner portion of the sapwood turns into heartwood as the tree ages.

As sapwood ripens into heartwood, the walls of any remaining living cells of xylem increasingly lignify, that is, harden to preserve cellulose. After they lignify, the xylem cells die. They lose their water content; oils, resins, tannins, and gums accumulate (Browse, 1988: 133). These latter materials make heartwood darker and more decay-resistant than sapwood. Heartwood is the wood extending from the pith, the soft-centred core of the tree, to the sapwood. Heartwood cells no longer participate in the life processes of the tree; however, they strengthen the tree and make it rigid if not exposed to air.

The radius of the heartwood increases with the age of the tree whereas that of the sapwood remains approximately the same (Browse, 1988: 134).

11

GROWTH PROCESSES IN POPLAR

Poplar as a Resource

Poplars are widely distributed throughout the northern hemisphere, chiefly in the temperate zone. They are fast growing, short-lived trees. The wood is light and soft; it has inconspicuous growth rings. It is used for veneer, plywood, lumber, boxes, small woodenware, and barrels for dry goods. As a pulpwood species, poplar ranks high among the hardwood species, and it is also the main resource for the fibreboard and particleboard industry (Farrar, 1995: 118).

Poplars have the widest range and greatest volume of any of the hardwood genera, yet only an insignificant part of the theoretical yield is harvested and utilized. In Europe and Asia, the genus has received much scientific study and has been intensively cultivated in many countries over a wide range of growing conditions for many years (Pfeiffer, 1978: 2).

In Canada, the general trend toward a greater utilization of hardwood species, together with reduced wood supplies in areas close to mills, is focussing attention on species such as poplar that have a capacity for

12

high yields on short rotations (Pfeiffer, 1978: 2).

Primary Growth

Importance to Vegetative Reproduction

Plantations of either native or hybrid species take advantage of primary growth characteristics of poplar by utilizing vegetative reproduction. Most poplar species propagate chiefly by means of suckers, or new stems, that arise from roots near the surface of the ground, a characteristic of great importance in regenerating poplar on cut-over or burned-over areas (Farrar, 1995: 118).

Another method of asexual or vegetative reproduction is stump and collar sprouts. If a stem is cut and part of the stump or root collar is left in the ground, new shoots will again appear. Depending on when the stems are cut, this process can happen more than once in the same growing season (Barkley, 1983: 3).

Factors Affecting Primary Growth

Poplar suckers originate singly or in clumps; the height of dominant shoots in a clump increase with an increase in the number of shoots per

13

clump. Height growth of trembling aspen (*Populus tremuloides*) suckers is initially faster than that of seedlings because of their already well-developed root system (Pfeiffer, 1978: 2).

Poplars have high moisture content. The average moisture content of trembling aspen in winter decreases in summer months. Since height growth occurs only in the apical meristem by the absorption of water and the moisture content of poplar is high, the best primary growth will take place on a moist site. For initial establishment of native and hybrid poplar species, sufficient moisture combined with well-aerated soil are basic requirements (Barkley, 1983: 8).

Secondary Growth

Secondary growth in poplar produces diameter faster as primary growth slows.

Factors Affecting Secondary Growth

A short rotation age is desirable because poplar is prone to decay as the trees get older (Armson and Smith, 1978: 21). A rotation age of between 30 and 50 years would offset this problem. However, poplar-based industries then have to make greater use of small logs if the problem of decay is to be overcome (Fayle et al., 1979: 19).

Insects and diseases that attack the stem are also a limiting factor on utilization of poplar and affect the quality of secondary growth. Some examples of these are borers, heart rot, cankers, and galls. Various species attack the xylem, cambium, phloem, and sapwood regions of the stem (Barkley, 1983: 29). Together, insects and diseases cause great differences between gross stand volume estimates and actual net merchantable volume.

Short Rotation Yields

Poplar has the capacity for high yields on short rotations (Pfeiffer, 1978: 26). Since yield is measured in units of volume, poplar producers look for acceptable diameter growth in a short period. Processing machinery for pulp and paper, particleboard, and fibreboard industries is specific for specific sizes of logs.

15

CONCLUSION

Summary

There are two major processes involved in the growth of stems in trees: primary and secondary growth, which result in the formation of annual rings and heartwood and sapwood in the tree with its increasing maturity.

Primary growth is responsible for growth in height whereas secondary growth causes an increase in diameter. Both these processes result in an increase in volume and thus potential merchantable value. Terminal growth is usually completed in the early part of the growing season, but lateral growth continues sometimes until frost. Primary tissues make up a relatively small amount of volume, and after secondary growth has proceeded for some time, these tissues play an insignificant role in the life of the tree.

With respect to poplar (*Populus*), primary growth is rapid in the various forms of vegetative reproduction: root suckering and stump sprouts. Sufficient moisture and well-aerated soil are two conditions that promote primary growth.

16

Secondary growth in poplar may produce adequate volumes for pulpwood and other uses on a short-rotation basis. Hybrid species are able to supply this volume successfully, although adapting existing processing machinery to small-diameter logs is an economic barrier. Because poplar is prone to decay and attacks from insects and disease with increasing age, good diameter growth in a short time period is doubly important.

Discussion

Mini-rotations

In the search to develop fast-growing trees with good-quality wood, federal and provincial research centres have crossed native and European species of poplar to produce hybrids with the following characteristics:

- frost-hardiness to utilize northern sites
- high survival rate and growth of root suckers
- good height and diameter growth in stump sprouts

These mini-rotation experiments to improve volume complement other experiments to develop a poplar crop that will be resistant to many diseases and insects. The new Ontario poplar clones (Anonymous, 1984) will increase the net merchantable volume and solve the problem of high cull rates.

Short-rotation Biomass Production

In eastern Ontario, for instance, hybrid poplar clones are producing annual dried biomass comparable to field crops (Pfeiffer, 1978: 4). Hybrid poplar for pulp yields biomass amounts similar to native poplar; however, it has slightly different physical properties that require chemical and physical adjustments in the pulping process (see Table 1).

Table 1. Pulp yields for hybrid poplar, native poplar, and mixed hardwoods (% harvested wood) (Pfeiffer, 1978: 5)

Wood Tested	Magnefite Pulps		Kraft Pulps	
	Unbleached	Bleached	Unbleached	Bleached
Hybrid cottonwood (a)	50.2	47.3	49.7	47.8
Hybrid cottonwood (b)	50.1	47.2	50.2	48.3
Hybrid aspen	51.9	49.6	51.3	49.6
Regular mill poplar	53.7	50.7	55.4	53.6
Dense mixed hardwoods	47.5	43.9	46.9	45.3

18

Experiments with hybrid poplar as a livestock feed suggest that ruminants can digest *in vitro* up to 47% of aspen species of poplar with appropriate treatment of the wood. Treated woods can contribute to dietary energy and roughage needs of ruminants. Using hybrid poplar as fuel is more problematic. Burning wood is inefficient and polluting. However, wood will produce methanol to blend with petroleum as a source of fuel (Pfeiffer, 1978: 5-18).

The aim is a continuous production on a yearly basis of adequate volume to meet demands for veneer, plywood, lumber, particleboard, and fibreboard, as our supplies of conifers diminish or become too expensive to harvest. This objective has been successful in some areas of Canada. As better hybrids and poplar management strategies develop, poplar farming will continue be a part of the Canadian forest industry.

19

REFERENCES

[Anonymous]. 1984. A guide to the identification of poplar clones in Ontario. Maple (ON): Ontario Tree Improvement and Forest Biomass Institute, Ontario Ministry of Natural Resources.

Armson KA, Smith JHG. 1978. Management of hybrid poplar. Case study 5, in Forest management in Canada. Canadian Forestry Service FMR-X-103. Ottawa, ON. 27 p.

Barkley BA. 1983. A silvicultural guide for hybrid poplar in Ontario. Toronto: Ontario Ministry of Natural Resources.

Browse PM. 1988. Plant propagation. New York: Simon & Schuster. 1st Fireside ed.

Farrar, JL. 1995. Trees in Canada. Markham (ON): Fitzhenry & Whiteside and Canadian Forest Service, Natural Resources Canada in cooperation with the Canada Communication Group-Publishing, Supply and Services Canada.

Fayle DCF, Zsuffa L, Anderson HW, editors. 1979. Poplar research, management and utilization in Canada. Proceedings of the North American Poplar Council Annual Meeting. Brockville, Ontario. 6-9 September 1977. Toronto: Ontario Ministry of Natural Resources.

Luger JH, Jarrell, RH. 2001. Botany of woody stems. Boston: Sevier.

Palmer CL. 1991. Short-rotation culture of populus and larix: a literature review. Sault Ste. Marie (ON): Canada-Ontario Forest Resource Development Agreement.

Pfeiffer WC. 1978. Economic potentials of hybrid poplar-based fibre production as an agricultural enterprise in eastern Ontario. Toronto: Ontario Ministry of Natural Resources

Shigo AL. 1994. Tree anatomy. Durham (NH): Shigo and Trees, Associates.

INDEX